Selective Laser Melting

Selective Laser Melting

Materials and Applications

Special Issue Editor

Prashanth Konda Gokuldoss

MDPI • Basel • Beijing • Wuhan • Barcelona • Belgrade

Special Issue Editor
Prashanth Konda Gokuldoss
Tallinn University of Technology
Estonia
Austrian Academy of Sciences
Austria

Editorial Office
MDPI
St. Alban-Anlage 66
4052 Basel, Switzerland

This is a reprint of articles from the Special Issue published online in the open access journal *Journal of Manufacturing and Materials Processing* (ISSN 2504-4494) from 2018 to 2020 (available at: https://www.mdpi.com/journal/jmmp/special_issues/SLM).

For citation purposes, cite each article independently as indicated on the article page online and as indicated below:

LastName, A.A.; LastName, B.B.; LastName, C.C. Article Title. *Journal Name* **Year**, *Article Number*, Page Range.

ISBN 978-3-03928-578-5 (Pbk)
ISBN 978-3-03928-579-2 (PDF)

© 2020 by the authors. Articles in this book are Open Access and distributed under the Creative Commons Attribution (CC BY) license, which allows users to download, copy and build upon published articles, as long as the author and publisher are properly credited, which ensures maximum dissemination and a wider impact of our publications.

The book as a whole is distributed by MDPI under the terms and conditions of the Creative Commons license CC BY-NC-ND.

Contents

About the Special Issue Editor vii

Konda Gokuldoss Prashanth
Selective Laser Melting: Materials and Applications
Reprinted from: *J. Manuf. Mater. Process.* **2020**, 4, 13, doi:10.3390/jmmp4010013 1

Wolfgang Schneller, Martin Leitner, Sebastian Springer, Florian Grün and Michael Taschauer
Effect of HIP Treatment on Microstructure and Fatigue Strength of Selectively Laser Melted AlSi10Mg
Reprinted from: *J. Manuf. Mater. Process.* **2019**, 3, 16, doi:10.3390/jmmp3010016 4

Altaf Ahmed, Arfan Majeed, Zahid Atta and Guozhu Jia
Dimensional Quality and Distortion Analysis of Thin-Walled Alloy Parts of AlSi10Mg Manufactured by Selective Laser Melting
Reprinted from: *J. Manuf. Mater. Process.* **2019**, 3, 51, doi:10.3390/jmmp3020051 13

Floriane Zongo, Antoine Tahan, Ali Aidibe and Vladimir Brailovski
Intra- and Inter-Repeatability of Profile Deviations of an AlSi10Mg Tooling Component Manufactured by Laser Powder Bed Fusion
Reprinted from: *J. Manuf. Mater. Process.* **2018**, 2, 56, doi:10.3390/jmmp2030056 28

Patrick Hartunian and Mohsen Eshraghi
Effect of Build Orientation on the Microstructure and Mechanical Properties of Selective Laser-Melted Ti-6Al-4V Alloy
Reprinted from: *J. Manuf. Mater. Process.* **2018**, 2, 69, doi:10.3390/jmmp2040069 42

Okanmisope Fashanu, Mario F. Buchely, Myranda Spratt, Joseph Newkirk, K. Chandrashekhara, Heath Misak and Michael Walker
Effect of SLM Build Parameters on the Compressive Properties of 304L Stainless Steel
Reprinted from: *J. Manuf. Mater. Process.* **2019**, 3, 43, doi:10.3390/jmmp3020043 55

Marios M. Fyrillas, Yiannos Ioannou, Loucas Papadakis, Claus Rebholz, Allan Matthews and Charalabos C. Doumanidis
Phase Change with Density Variation and Cylindrical Symmetry: Application to Selective Laser Melting
Reprinted from: *J. Manuf. Mater. Process.* **2019**, 3, 62, doi:10.3390/jmmp3030062 70

About the Special Issue Editor

Prashanth Konda Gokuldoss (Prof.) is the Head of the Additive Manufacturing Laboratory and Professor in Additive Manufacturing at the Department of Mechanical and Industrial Engineering, Tallinn University of Technology, Tallinn, Estonia. He is a guest scientist at the Erich Schmid Institute of Materials Science, Austrian Academy of Science, Leoben, Austria and an Adjunct Professor at the department of CBCMT, School of Engineering, Vellore Institute of Technology, Vellore, India. He received a Ph.D. from the Technical University Dresden, Germany (2014), and conducted postdoctoral research at the Leibniz Institute of Solid State and Materials Research (IFW) Dresden, Germany. He has also worked as a R&D Engineer (Sandvik, Sweden), Senior Scientist (Erich Schmid Institute of Materials Science, Austrian Academy of Science, Leoben, Austria), and Associate Professor (Norwegian University of Science and Technology, Gjøvik, Norway) before taking a Full Professorship at the Tallinn University of Technology, Tallinn, Estonia. His present research is focused on, but not limited to, additive manufacturing (alloys, process, and product development), fabrication of meta-stable materials, powder metallurgy, light materials, solidification, and biomaterials. He has published over 125 peer reviewed journal papers with an H-index of 32 (Google scholar). A multiple award winner, he actively collaborates with and visits China, India, the USA, Austria, Poland, Norway, Germany, Spain, Taiwan, South Korea, and Iran.

Journal of
Manufacturing and Materials Processing

Editorial

Selective Laser Melting: Materials and Applications

Konda Gokuldoss Prashanth [1,2,3]

1. Department of Mechanical and Industrial Engineering, Tallinn University of Technology, Ehitajate Tee 5, 19086 Tallinn, Estonia; kgprashanth@gmail.com; Tel.: +372-5452-5540
2. Erich Schmid Institute of Materials Science, Austrian Academy of Science, Jahnstrasse 12, A-8700 Leoben, Austria
3. CBCMT, School of Mechanical Engineering, Vellore Institute of Technology, Vellore, Tamil Nadu 632014, India

Received: 17 February 2020; Accepted: 17 February 2020; Published: 18 February 2020

Additive manufacturing (AM) is one of the emerging manufacturing techniques of immense engineering and scientific importance and is regarded as the technique of the future [1–3]. AM can fabricate any kind of material, including metals, polymers, ceramics, composites, etc. Selective laser melting (SLM), also known as the laser-based powder bed fusion process (LPBF), is the most widely used AM techniques that can fabricate a wide variety of materials, including Al-based [4–6], Fe-based [7–10], Ti-based [11–13], Co-based [14–16], Cu-based [17–19] and Ni-based alloys [20–22], etc. Similar to any AM processes, the SLM/LPBF process also offers several advantages, like added functionality, near-net-shape fabrication with minimal or no post-processing, shorter lead-time, offer intricacy for free, etc. [23–25]. The SLM process has its applications in the aerospace, automobile, oil refinery, marine, construction, food and jewelry industries, etc. [26–28]. However, there exist some shortcomings in the SLM field, which are (a) SLM-based alloy development [29], (b) the premature failure of materials, even though improved properties are observed [30], (c) process innovation and development, (d) structure-property correlation and (e) numerical simulations, etc.

Accordingly, the present Special Issue (book) focuses on the two main aspects: materials and applications. Alloy design and development that suits the specific process conditions is essential, rather than using the conventionally designed/available materials. The application spectrum is getting wider day by day, hence the need for our attention. Overall, six articles are published under this Special Issue, with the following themes:

- AlSi10Mg alloy focusing on microstructure and fatigue properties with the influence of HIP process [31], dimensional and distortion analysis of thin walled parts [32] and intra- and inter-repeatability of profile deviations in tooling components (3 articles) [33].
- Ti6Al4V—effect of build orientation with microstructure-property correlations (1 article) [34].
- 304L—correlation between build parameters and compressive properties (1 article) [35] and
- Finally, phase change with density variation and cylindrical symmetry—applications to SLM (1 article) [36].

The outcome of the Special Issue suggests that research is thriving in the field of SLM, especially in microstructure and property correlations. The present Special Issue is interesting particularly because it covers different materials, including AlSi10Mg, Ti6Al4V and 304L stainless steel and gives an overview of microstructure-property correlation in this field.

Finally, we would like to thank all the contributing authors for their excellent contributions to this Special Issue, to the reviewers for constructively improving the quality of the Special Issue and to the *JMMP* staff for giving us the opportunity to host this Special Issue and for the timely publication of the articles.

Funding: European Regional Development Fund funded the research through project MOBERC15.

Conflicts of Interest: The author declares no conflict of interest.

References

1. Oliveria, J.P.; Santos, T.G.; Miranda, R.M. Revisiting fundamental welding concepts to improve additive manufacturing: From theory to practice. *Prog. Mater. Sci.* **2020**, *107*, 100590. [CrossRef]
2. Prashanth, K.G.; Scudino, S.; Eckert, J. Defining the tensile properties of Al-12Si parts produced by selective laser melting. *Acta Mater.* **2017**, *126*, 25–35. [CrossRef]
3. Herzong, D.; Seyda, V.; Wycisk, E.; Emmelmann, C. Addiitive manufacturing of metals. *Acta Mater.* **2016**, *117*, 371–392. [CrossRef]
4. Prashanth, K.G.; Scudino, S.; Eckert, J. Tensile properties of Al-12Si fabricated by selective laser melting (SLM) at different temperatures. *Technologies* **2016**, *4*, 38. [CrossRef]
5. Prashanth, K.G.; Shakur Shahabi, H.; Attar, H.; Srivastava, V.C.; Ellendt, N.; Uhlenwinkel, V.; Eckert, J.; Scudino, S. Production of high strength $Al_{85}d_8Ni_5Co_2$ alloy by selective laser melting. *Addit. Manuf.* **2015**, *6*, 1–5. [CrossRef]
6. Prashanth, K.G.; Scudino, S.; Klauss, H.-H.; Surreddi, K.B.; Löber, L.; Wang, Z.; Chaubey, A.K.; Kühn, U.; Eckert, J. Microstructure and mechanical properties of Al-12Si produced by selective laser melting: Effect of heat treatment. *Mater. Sci. Eng. A* **2014**, *590*, 153–160. [CrossRef]
7. Suryawanshi, J.; Prashanth, K.G.; Ramamurty, U. Mechanical behavior of selective laser melted 316L stainless steel. *Mater. Sci. Eng. A* **2017**, *696*, 113–121. [CrossRef]
8. Suryawanshi, J.; Prashanth, K.G.; Ramamurty, U. Tensile, fracture and fatigue crack growth properties of a 3D printed maraging steel through selective laser melting. *J. Alloys Compd.* **2017**, *725*, 355–364. [CrossRef]
9. Jung, H.Y.; Choi, S.J.; Prashanth, K.G.; Stoica, M.; Scudino, S.; Yi, S.; Kühn, U.; Kim, D.H.; Kim, K.B.; Eckert, J. Fabrication of Fe-based bulk metallic glass by selective laser melting: A parameter study. *Mater. Des.* **2015**, *86*, 703–708. [CrossRef]
10. Prashanth, K.G.; Löber, L.; Klauss, H.-J.; Kühn, U.; Eckert, J. Characterization of 316L steel cellular dodecahedron structures produced by selective laser melting. *Technologies* **2016**, *4*, 34. [CrossRef]
11. Attar, H.; Prashanth, K.G.; Chaubey, A.K.; Calin, M.; Zhang, L.C.; Scudino, S.; Eckert, J. Comparison of wear properties of commercially pure titanium prepared by selective laser melting and casting processes. *Mater. Lett.* **2015**, *142*, 38–41. [CrossRef]
12. Schwab, H.; Prashanth, K.G.; Löber, L.; Kühn, U.; Eckert, J. Selective laser melting of Ti-45Nb alloy. *Metals* **2015**, *5*, 686–694. [CrossRef]
13. Attar, H.; Löber, L.; Funk, A.; Calin, M.; Zhang, L.C.; Prashanth, K.G.; Scudino, S.; Zhang, Y.S.; Eckert, J. Mechanical behavior of porous commercially pure Ti and Ti-TiB composite materials manufactured by selective laser melting. *Mater. Sci. Eng. A* **2015**, *625*, 350–356. [CrossRef]
14. Song, C.; Zhang, M.; Yang, Y.; Wang, D.; Jia-Kuo, Y. Morphology and properties of CoCrMo parts fabricated by selective laser melting. *Mater. Sci. Eng. A* **2018**, *713*, 206–213. [CrossRef]
15. Hedberg, Y.S.; Qian, B.; Shen, Z.; Virtanen, S.; Wallinder, I.O. In vitro biocompatibility of CoCrMo dental alloys fabricated by selective laser melting. *Dent. Mater.* **2014**, *30*, 525–534. [CrossRef]
16. Tonelli, L.; Fortunato, A.; Ceschini, L. CoCr alloy processed by selective laser melting (SLM): Effect of laser energy density on microstructure, surface morphology, and hardness. *J. Manuf. Process.* **2020**, *52*, 106–119. [CrossRef]
17. Scudino, S.; Unterdoerfer, C.; Prashanth, K.G.; Attar, H.; Ellendt, N.; Uhlenwinkel, V.; Eckert, J. Additive manufacturing of Cu-10Sn bronze. *Mater. Lett.* **2015**, *156*, 202–204. [CrossRef]
18. Wang, J.; Zhou, X.L.; Li, J.; Brochu, M.; Zhao, Y.F. Microstructures and properties of SLM manufactured Cu-15Ni-8Sn alloy. *Addit. Manuf.* **2020**, *31*, 100921. [CrossRef]
19. Murray, T.; Thomas, S.; Wu, Y.; Neil, W.; Hutchinson, C. Selective laser melting of nickel aluminium bronze. *Addit. Manuf.* **2020**, *X*, 101122. [CrossRef]
20. Ren, D.C.; Zhang, H.B.; Liu, Y.J.; Li, S.J.; Jin, W.; Wang, R.; Zhang, L.C. Microstructure and properties of equiatomic Ti-Ni alloy fabricated by selective laser melting. *Mater. Sci. Eng. A* **2020**, *771*, 138586. [CrossRef]
21. Zhang, B.; Xi, M.; Tan, Y.T.; Wei, J.; Wang, P. Pitting corrosion of SLM Inconel 718 sample under surface and heat treatments. *Appl. Surf. Sci.* **2019**, *490*, 556–567. [CrossRef]
22. Zhang, Q.; Hao, S.; Liu, Y.; Xiong, Z.; Guo, W.; Yang, Y.; Ren, Y.; Cui, L.; Ren, L.; Zhang, Z. The microstructure of a selective laser melting (SLM)-fabricated NiTi shape memory alloy with superior tensile property and shape memory recoverability. *Appl. Mater. Today* **2020**, *19*, 100547. [CrossRef]

23. Maity, T.; Chawke, N.; Kim, J.T.; Eckert, J.; Prashanth, K.G. Anisotropy in local microstructure – Does it affect the tensile properties of the SLM sample? *Manuf. Lett.* **2018**, *15*, 33–37. [CrossRef]
24. Prashanth, K.G.; Eckert, J. Formation of metastble cellular microstructures in selective laser melted alloys. *J. Alloys Compd.* **2017**, *707*, 27–34. [CrossRef]
25. Ma, P.; Jia, Y.; Prashanth, K.G.; Scudino, S.; Yu, Z.; Eckert, J. Microstructure and phase formation in Al-20Si-5Fe-3Cu-1Mg synthesized by selective laser melting. *J. Alloys Compd.* **2016**, *657*, 430–435. [CrossRef]
26. Prashanth, K.G.; Kolla, S.; Eckert, J. Additive manufacturing processes: Selective laser melting, electron beam melting and binder jetting—Selection guidelines. *Materials* **2017**, *10*, 672. [CrossRef]
27. Wang, P.; Li, H.C.; Prashanth, K.G.; Eckert, J.; Scudino, S. Selective laser melting of Al-Zn.Mg-Cu: Heat treatment, microstructure and mechanical properties. *J. Alloys Compd.* **2017**, *707*, 287–290. [CrossRef]
28. Xi, L.X.; Zhang, H.; Wang, P.; Li, H.C.; Prashanth, K.G.; Lin, K.J.; Kaban, I.; Gu, D.D. Comparative investigation of microstructure, mechanical properties and strengthening mechanisms of Al-12Si/TiB$_2$ fabricated by selective laser melting and hot pressing. *Ceram. Int.* **2018**, *44*, 17635–17642. [CrossRef]
29. Prashanth, K.G. Design of next-generation alloys for additive manufacturing. *Mater. Des. Process. Commun.* **2019**, *1*, e50. [CrossRef]
30. Prashanth, K.G. Work hardening in selective laser melted Al-12Si alloy. *Mater. Des. Process. Commun.* **2019**, *1*, e46. [CrossRef]
31. Fyrillas, M.M.; Ioannou, Y.; Papadakis, L.; Rebholz, C.; Matthews, A.; Doumanidis, C.C. Phase change with density variation and cylindrical symmetry: Application to selective laser melting. *J. Manuf. Mater. Process.* **2019**, *3*, 62. [CrossRef]
32. Fashanu, O.; Buchley, M.F.; Spratt, M.; Newkirk, J.; Chandrashekhara, K.; Misak, H.; Walker, M. Effect of SLM build parameters on the compressive properties of 304L stainless steel. *J. Manuf. Mater. Process.* **2019**, *3*, 43. [CrossRef]
33. Hartunian, P.; Eshragi, M. Effect of build orientation on the microstructure and mechanical properties of selective laser melted Ti-6Al-4Valloy. *J. Manuf. Mater. Process.* **2018**, *2*, 69. [CrossRef]
34. Zongo, F.; Tahan, A.; Aidibe, A.; Brailovski, V. Intra- and Inter-repeatability of profile deviations of an AlSi10Mg tooling component manufactured by laser powder bed fusion. *J. Manuf. Mater. Process.* **2018**, *2*, 56. [CrossRef]
35. Ahmad, A.; Majeed, A.; Atta, A.; Jia, G. Dimensional quality and distortion analysis of thing-walled alloy parts of AlSi10Mg manufactured by selective laser melting. *J. Manuf. Mater. Process.* **2019**, *3*, 51. [CrossRef]
36. Schneller, W.; Leitner, M.; Springer, S.; Gruen, F.; Taschauer, M. Effect of HIP treatment on microstructure and fatigue strength of selectively laser melted AlSi10Mg. *J. Manuf. Mater. Process.* **2019**, *3*, 16. [CrossRef]

© 2020 by the author. Licensee MDPI, Basel, Switzerland. This article is an open access article distributed under the terms and conditions of the Creative Commons Attribution (CC BY) license (http://creativecommons.org/licenses/by/4.0/).

Article

Effect of HIP Treatment on Microstructure and Fatigue Strength of Selectively Laser Melted AlSi10Mg

Wolfgang Schneller [1,*], Martin Leitner [1], Sebastian Springer [1], Florian Grün [1] and Michael Taschauer [2]

[1] Department Product Engineering, Chair of Mechanical Engineering, Montanuniversität Leoben, 8700 Leoben, Austria; martin.leitner@unileoben.ac.at (M.L.); sebastian.springer@unileoben.ac.at (S.S.); florian.gruen@unileoben.ac.at (F.G.)
[2] Pankl Systems Austria GmbH, 8605 Kapfenberg, Austria; michael.taschauer@pankl.com
* Correspondence: wolfgang.schneller@unileoben.ac.at; Tel.: +43-3842-402-1451

Received: 15 December 2018; Accepted: 29 January 2019; Published: 1 February 2019

Abstract: This study shows the effect of hot isostatic pressing (HIP) on the porosity and the microstructure, as well as the corresponding fatigue strength of selectively-laser-melted (SLM) AlSi10Mg structures. To eliminate the influence of the as-built surface, all specimens are machined and exhibit a polished surface. To highlight the effect of the HIP treatment, the HIP specimens are compared to a test series without any post-treatment. The fatigue characteristic is evaluated by tension-compression high cycle fatigue tests under a load stress ratio of $R = -1$. The influence of HIP on the microstructural characteristics is investigated by utilizing scanning electron microscopy of micrographs of selected samples. In order to study the failure mechanism and the fatigue crack origin, a fracture surface analysis is carried out. It is found that, due to the HIP process and subsequent annealing, there is a beneficial effect on the microstructure regarding the fatigue crack propagation, such as Fe-rich precipitates and silicon agglomerations. This leads, combined with a significant reduction of global porosity and a decrease of micro pore sizes, to an improved fatigue resistance for the HIPed condition compared to the other test series within this study.

Keywords: additive manufacturing; SLM; AlSi10Mg; fatigue strength; HIP; porosity

---

1. Introduction

Additive manufacturing (AM) offers the possibility to manufacture complexly-shaped and topographically-optimized components [1–5]. Therefore, powder bed-based AM is contemplated to find application in various fields such as aviation, automotive, and biomedical engineering [6]. Estimations state that 55% of all failures in aeronautic engineering and, generally speaking, about 90% of all engineering failures are caused by a fatigue-related damage mechanism [7,8]. Hence, it is of upmost importance to investigate and understand the fracture mechanisms and fatigue characteristics, to assess properly, as well as safely the material qualifications. It is crucial to take account of the interaction between the microstructure, internal defects, and fatigue resistance [9,10].

Inner imperfections like unmolten areas or bonding errors between melt-pool borders and pores are mostly responsible for fatigue failures concerning AM components. It is necessary to control the distribution and extension of such cavities, as they are preferable spots for fatigue crack initiation [11,12]. Given the fact that in the case of cast aluminum alloys, hot isostatic pressing (HIP) significantly decreases the volume fraction of porosity with only minor changes of microstructural features, leading to a considerable increase of fatigue strength, an appropriate post-treatment may be beneficial to AM parts, as well [13–16]. One can find that due to the extremely fine microstructure of selectively-laser-melted (SLM) parts, an HIP treatment above the solubility temperature of AlSi10Mg leads to microstructural coarsening because of the dissolving of grain boundaries. This results in a reduced fatigue resistance, although the porosity is significantly lower [8,17]. To take advantage of the beneficial effect of HIP on the porosity, the changes within the microstructure cause the necessity of quenching and a subsequent age hardening process to counteract these negative effects [18]. The exact HIP parameter was determined incorporating the knowledge of the specimen manufacturer with the aim of reducing the amount of porosity in order to improve the fatigue behavior.

For this reason, the fatigue strength of the HIP-treated specimen at a commonly-used temperature for solution annealing followed by low temperature annealing as heat treatment was investigated. Besides their fatigue resistance, the local material properties, such as porosity and microstructure, were analyzed and compared to specimens without any post-treatment, denoted as the as-built condition.

2. Materials and Methods

The chemical composition of the utilized AlSi10Mg powder, shown in Table 1, is given by the manufacturer specification and corresponds to the standard DIN EN 1706:2010 [19].

Table 1. Chemical composition of the additive manufacturing (AM) powder by weight %.

Material	Si	Fe	Cu	Mn	Mg	Al
AlSi10Mg	9.0–11.0	0.55	0.05	0.45	0.20–0.45	Balance

Specimens were fabricated using an EOS M290 system with a Yb fiber laser, a power of 400 W, and a beam diameter of 100 µm. All specimens were built in the vertical direction with a certain machining allowance in order to remove subsequently the as-built surface and eliminate surface-related effects. The structures were manufactured according to the standard parameter set given by the system and powder manufacturer EOS. Following the built process, hot isostatic pressing was performed applying a temperature higher than 500 °C and a pressure of above 100 MPa with a holding time of at least two hours followed by quenching under constant pressure. Low temperature annealing over a certain time period was conducted afterwards. Subsequent to the heat treatment, the specimens were processed to the final geometry by turning and polishing. A CAD drawing with the detailed specimen geometry and dimensions is shown in Figure 1. The shape of the specimens was designed to show a homogeneous stress distribution over the cross-section with a stress concentration factor as low as possible due to the narrowing within the testing section, corresponding to no common standard.

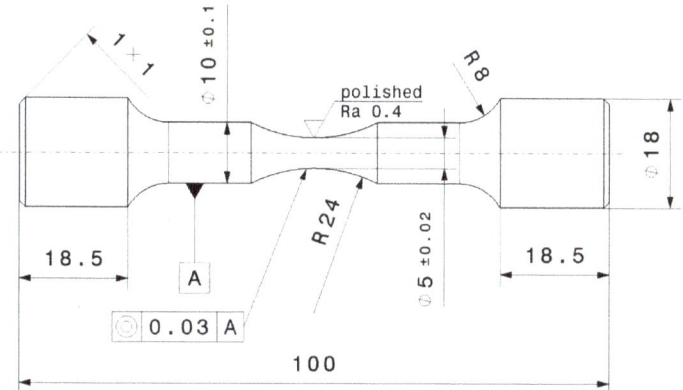

Figure 1. CAD drawing of the specimen geometry for the high cycle fatigue test.

The specimens are fatigue tested at a load stress ratio of R = −1 on a RUMUL Mikrotron resonant testing rig with a frequency of about 106 Hz. Collets were used for gripping in order to clamp the specimen at both ends. The abort criterion was defined either as total fracture or as run-out at 1×10^7 load cycles. Run-outs were reinserted at higher stress levels to obtain more data in the finite life regime, conservatively assuming pre-damaging at stress levels lower than the endurance limit [20]. For each test series, respectively with and without HIP treatment, nine specimens were manufactured and tested.

3. Results and Discussion

3.1. Effect of HIP Treatment on the Microstructure

HIP treatment at high temperature with considerably high pressure leads to significant microstructural differences compared to the as-built condition; hence, the effect on the material was investigated in detail. To characterize the microstructure after HIP and heat treatment, SEM images, taken with a Carl Zeiss EVO MA 15 microscope, of the post-processed condition were evaluated. In Figure 2, one can clearly see Fe-rich precipitates and Si particles, which were also detected in [21]. Silicon crystals were precipitated at the grain boundaries during the HIP treatment above the solubility temperature, and they grew to their respective size during low temperature annealing [22–25]. Microstructural features like silicon agglomerations and needle-shaped, Fe-rich precipitates obstructed a propagating fatigue crack and, therefore, generally improved the resistance against fatigue crack growth. Such microstructures favor crack deflection and energy dissipation at the crack tip. Hence, the long crack growth was decelerated, whereby the fatigue strength was enhanced [17,26].

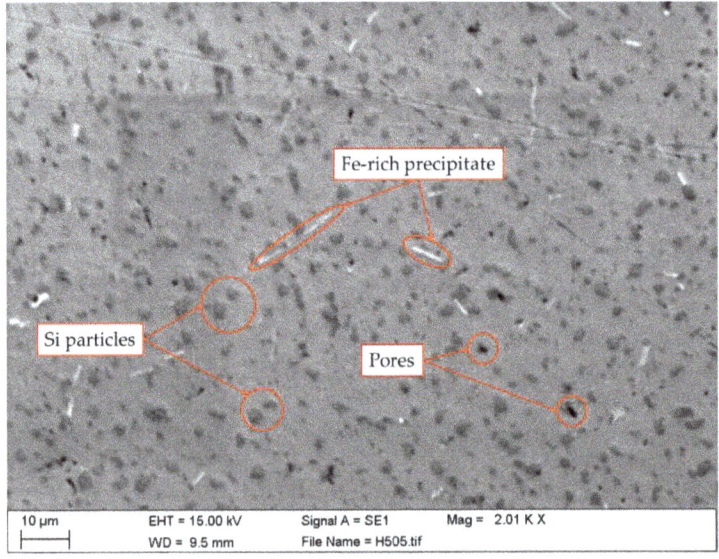

Figure 2. Microstructure after HIP and subsequent heat treatment.

Comparing the microstructure of the as-built condition (Figure 3a) to the microstructure after the post-treatment (Figure 3b,c), appreciable differences regarding the porosity we observed. For that reason, these figures have the same magnification and scale. A larger magnification image is depictured in Figure 3d, which reveals a circular shape of the observed micro-porosity. One can see that the amount of porosity and the maximum extension of pores have significantly decreased. Additionally, after the post-treatment, melt-pool boundaries completely vanished. The aforementioned Fe-rich precipitates and Si-crystals were formed within the microstructure. Throughout the annealing, the Si-particles grew at Si-rich cellular boundaries, and finally, grain boundaries were no longer clearly visible at this stage due to the heat influence [23]. The comparison between backscatter images before (Figure 3a) and after (Figure 3b) HIP treatment highlights this microstructural change.

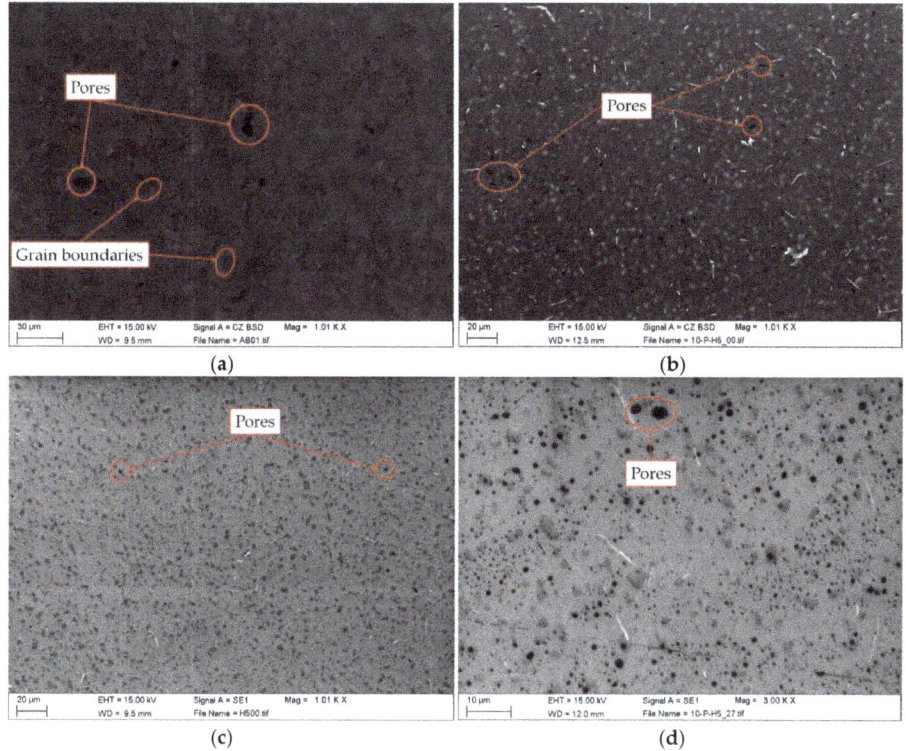

Figure 3. Microstructure (a) before and (b–d) after post-treatment.

3.2. Fatigue Tests

The fatigue test results are presented in Figure 4. The dashed line with square marks represents the data for the as-built series. The full line with triangle markings shows the data for the HIP condition. Within the finite life region, the specimen was tested at several load levels with a certain incrementation. The evaluation of the SN-curve in the finite life region is based on the ASTM E739 standard [27]. The high cycle fatigue strength at 1×10^7 load-cycles was statistically evaluated by applying the arcsin$\sqrt{P}$-transformation procedure given in [28]. Run-outs were reinserted at higher stress levels in order to obtain additional data within the finite life region. The results were normalized to the nominal ultimate tensile strength (UTS) of the additively-manufactured material without any post-treatment, given by the powder manufacturer [29]. The peak load level was set at about 35% of the UTS, which was well below the yield strength according to the powder manufacturer, to ensure testing within the linear-elastic region of the material and obtain reasonable results regarding endured load cycles. The results revealed that the HIP test series provided an increase of the high cycle fatigue strength of about 14% considering a survival probability of $P_S = 50\%$. The scatter band between 10% and 90% survival probability, referring to the stress amplitude, minorly decreased for the HIP condition compared to the as-built condition. Furthermore, the slope in the finite life region was less steep for the HIP condition. The fatigue test results are summarized in Table 2.

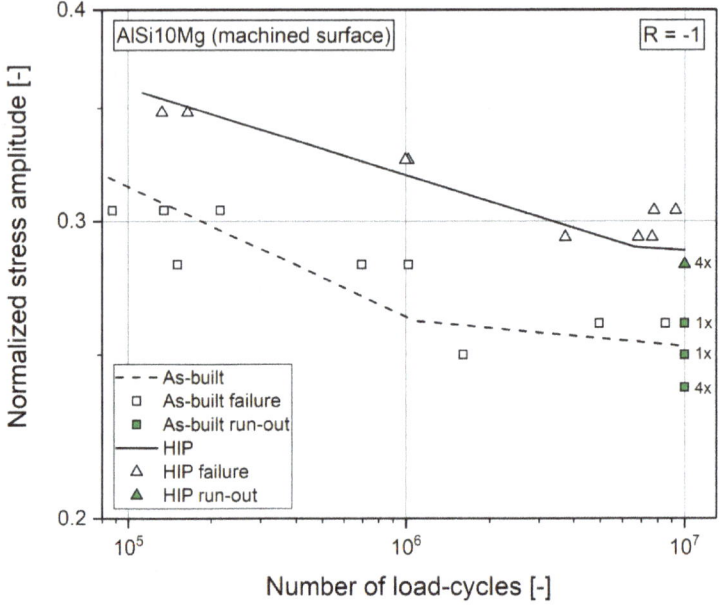

Figure 4. SN-curves for the as-built and HIP condition.

Table 2. Statistically evaluated SN-curve parameters for both test series.

Condition	Normalized Fatigue Strength ($P_S = 50\%$)	Difference	Slope in the Finite Life Region	Scatter Band in the Finite Life Region
As-built	0.253	Basis	12.99	1:1.15
HIP-treated	0.288	+14%	19.37	1:1.06

3.3. Metallographic and Fracture Surface Analysis

In order to evaluate the decrease in porosity, the average maximum pore extension, as well as the equivalent circle pore diameter, several micrographs of the two conditions were investigated. Figure 5a shows an example of the as-built condition, whereas Figure 5b is taken from the microsection of an HIP-treated specimen. All pictures of micrographs and fracture surfaces were recorded with a KEYENCE VHX-5000 light optical digital microscope. The microsections were prepared only by polishing and received no additional etching. Dependent on the polished surface and the image post-processing, different lighting options and angles were necessary. This was the reason why the as-built specimen in Figure 5a (ring-lighting) appears blue and shows a different texture, e.g., visible melting tracks and laser scanning strategy, than the HIP sample in Figure 5b (coaxial lighting). In order to determine the amount of porosity, image processing tools were utilized. At first, the images were converted to binary pictures with a certain threshold to ensure that the microsection of the specimen area appeared white while pores appeared black. Secondly, the embedding material was subtracted from the image. In the end, the separated pores, as well as the porosity, which is the ratio of specimen area to pore area, could easily be evaluated. The outcome is presented in Figure 6a–c and summarized in Table 3. The results were again normalized to the as-built condition to highlight the differences between the two test series. The results maintained that the HIP samples possessed a significant lower level of porosity (−64%), a decreased maximum pore extension (−22%), as well as an equivalent circle diameter (−11%).

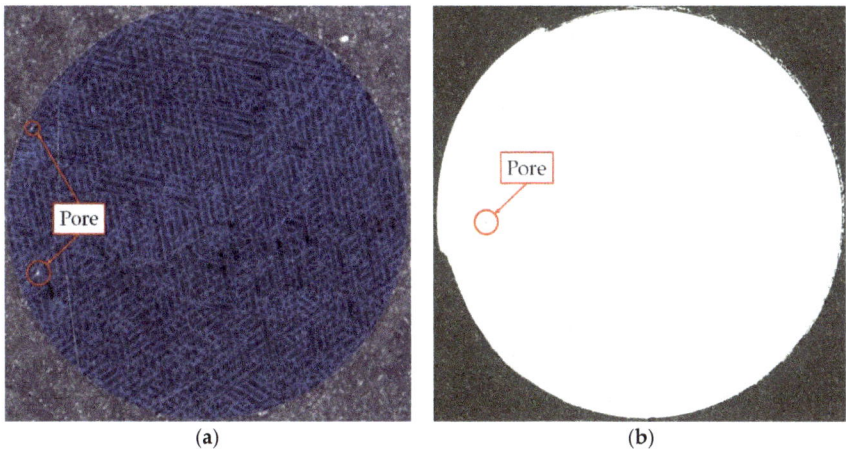

Figure 5. Micrograph of an (**a**) as-built and (**b**) HIP sample.

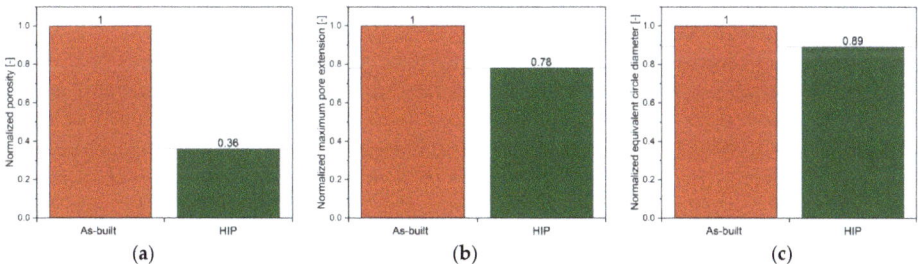

Figure 6. Difference in (**a**) porosity, (**b**) maximum pore extension, and (**c**) equivalent circle pore diameter between the as-built and HIP series.

Table 3. Summary of the porosity and pore size characteristics between the as-built and HIP condition.

Condition	Normalized Amount of Porosity	Normalized Maximum Pore Extension	Normalized Equivalent Circle Diameter
As-built	1.00 (Basis)	1.00 (Basis)	1.00 (Basis)
HIP-treated	0.36 (−64%)	0.78 (−22%)	0.89 (−11%)

To characterize the crack-initiating defect, a fracture surface analysis for each tested specimen was carried out. A fractured surface of the as-built specimen is presented in Figure 7a. The surface is visually differentiated into two sections, the oscillating crack growth regime and the burst fractured area. The defect, which was responsible for the failure, can be easily identified and evaluated. In every investigated fractured surface for the as-built condition, a pore was failure critical. An example with a marked and measured pore is given in Figure 7b. The size and location of the failure causing imperfection was one determining factor for the fatigue strength of the material; see also [30,31]. Therefore, an evaluation of the defect size was necessary to compare and to assess the fatigue strength of the two investigated conditions.

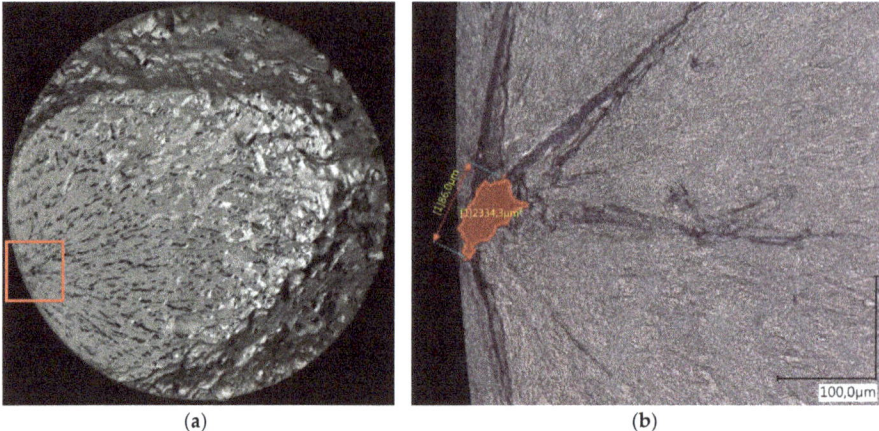

Figure 7. (a) Fracture surface of an as-built specimen; (b) size measurement of failure-critical defect.

A fracture surface for the post-processed condition (two-dimensional image with in depth focus) is displayed in Figure 8a. As pointed out for the as-built condition, the fracture surface is again separated into two different zones. The crack origin can be found within the fatigue fracture area, since the fine structured area points towards the crack initiation site. The fracture surface analysis for the HIP specimens revealed a different failure mechanism compared to the as-built ones. Due to the remarkable decrease in porosity, cavities were no longer responsible for fatigue crack initiation, but rather microstructural features such as silicon-rich phases. In Figure 8b, one can identify the debonding of Si-crystals as the failure origin; see also [26]. The crack initiated near the subsurface at all tested samples, for the HIP condition, as well as for the as-built condition. In almost every case, no evidence of pores could be found near the crack origin.

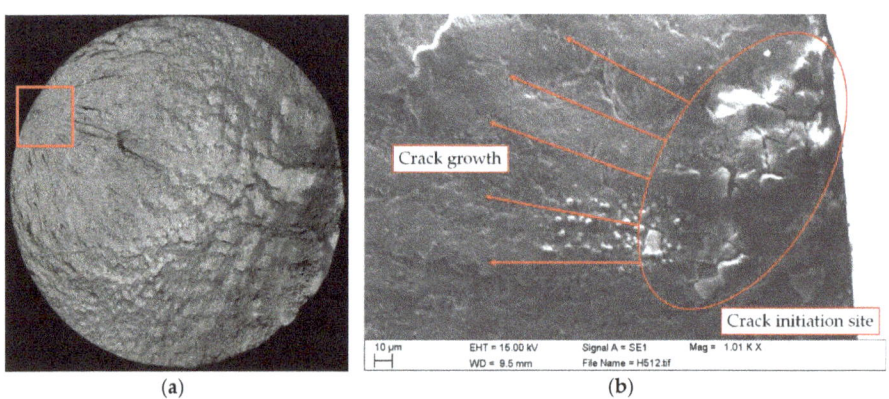

Figure 8. (a) Fracture surface of an HIP specimen; (b) failure-critical, microstructural inhomogeneity.

4. Conclusions

Based on the results presented in this paper, a beneficial effect on the fatigue strength of an HIP treatment above the solubility temperature with subsequent low temperature annealing can be observed for the additively-manufactured AlSi10Mg material. Concerning the microstructure, there was a significant decrease in porosity by 64%, maximum pore extension by 22%, and equivalent circle diameter by 11%. Because of the heat influence, melt-pool boundaries were dissolved, and grain boundaries were no longer visible due to the growth of Si-precipitates at the cellular boundaries.

After finishing the post-treatment, silicon agglomerations, as well as needle-shaped, iron-rich intermetallic phases were formed. These precipitates caused a deceleration of the crack growth due to the interference of the crack front at these microstructural features. Such a microstructure generally improves the resistance against fatigue crack growth since the propagation of the crack is obstructed. In summary, it was observed that the changes of the microstructure due to the application of the post-treatment contributed to an enhanced fatigue strength.

In addition, a change of the failure mechanism was also detected. For the as-built condition, pores were the decisive defect type. On the contrary, intermetallic inhomogeneities provoked the failure for the HIP condition. The crack initiation site is found in every case within the surface near region, independent of the failure mode. The combination of the microstructural changes consequently influenced the crack initiation, as well as the propagation behavior, leading to an improvement of 14% of the high cycle fatigue strength at a survival probability of 50% by the applied post-treatment.

Author Contributions: Conceptualization, W.S. and M.L.; methodology, W.S and M.L.; validation, W.S. and M.L.; formal analysis, W.S.; investigation, W.S. and S.S.; resources, W.S.; data curation, W.S. and S.S.; writing, original draft preparation, W.S.; writing, review and editing, W.S. and M.L.; visualization, W.S.; supervision, M.L.; project administration, M.L. and F.G.

Conflicts of Interest: The funders had no role in the design of the study; in the collection, analyses, or interpretation of data; in the writing of the manuscript; nor in the decision to publish the results.

References

1. Harun, W.; Kamariah, M.; Muhamad, N.; Ghani, S.; Ahmad, F.; Mohamed, Z. A review of powder additive manufacturing processes for metallic biomaterials. *Powder Technol.* **2018**, *327*, 128–151. [CrossRef]
2. Hedayati, R.; Hosseini-Toudeshky, H.; Sadighi, M.; Mohammadi-Aghdam, M.; Zadpoor, A.A. Computational prediction of the fatigue behavior of additively manufactured porous metallic biomaterials. *Int. J. Fatigue* **2016**, *84*, 67–79. [CrossRef]
3. Huynh, L.; Rotella, J.; Sangid, M.D. Fatigue behavior of IN718 microtrusses produced via additive manufacturing. *Mater. Des.* **2016**, *105*, 278–289. [CrossRef]
4. Leary, M.; Mazur, M.; Elambasseril, J.; McMillan, M.; Chirent, T.; Sun, Y.; Qian, M.; Easton, M.; Brandt, M. Selective laser melting (SLM) of AlSi12Mg lattice structures. *Mater. Des.* **2016**, *98*, 344–357. [CrossRef]
5. Watson, J.K.; Taminger, K. A decision-support model for selecting additive manufacturing versus subtractive manufacturing based on energy consumption. *J. Clean. Prod.* **2018**, *176*, 1316–1322. [CrossRef]
6. Herzog, D.; Seyda, V.; Wycisk, E.; Emmelmann, C. Additive manufacturing of metals. *Acta Mater.* **2016**, *117*, 371–392. [CrossRef]
7. Campbell, G.; Lahey, R. A survey of serious aircraft accidents involving fatigue fracture. *Int. J. Fatigue* **1984**, *6*, 25–30. [CrossRef]
8. Uzan, N.E.; Shneck, R.; Yeheskel, O.; Frage, N. Fatigue of AlSi10Mg specimens fabricated by additive manufacturing selective laser melting (AM-SLM). *Mater. Sci. Eng. A* **2017**, *704*, 229–237. [CrossRef]
9. Aboulkhair, N.T.; Maskery, I.; Tuck, C.; Ashcroft, I.; Everitt, N.M. Improving the fatigue behaviour of a selectively laser melted aluminium alloy: Influence of heat treatment and surface quality. *Mater. Des.* **2016**, *104*, 174–182. [CrossRef]
10. Domfang Ngnekou, J.N.; Nadot, Y.; Henaff, G.; Nicolai, J.; Kan, W.H.; Cairney, J.M.; Ridosz, L. Fatigue properties of AlSi10Mg produced by Additive Layer Manufacturing. *Int. J. Fatigue* **2019**, *119*, 160–172. [CrossRef]
11. Buffiere, J.-Y. Fatigue Crack Initiation and Propagation from Defects in Metals: Is 3D Characterization Important? *Procedia Struct. Integr.* **2017**, *7*, 27–32. [CrossRef]
12. Domfang Ngnekou, J.N.; Nadot, Y.; Henaff, G.; Nicolai, J.; Ridosz, L. Influence of defect size on the fatigue resistance of AlSi10Mg alloy elaborated by selective laser melting (SLM). *Procedia Struct. Integr.* **2017**, *7*, 75–83. [CrossRef]
13. Ceschini, L.; Morri, A.; Sambogna, G. The effect of hot isostatic pressing on the fatigue behaviour of sand-cast A356-T6 and A204-T6 aluminum alloys. *J. Mater. Process. Technol.* **2008**, *204*, 231–238. [CrossRef]

14. Lee, M.H.; Kim, J.J.; Kim, K.H.; Kim, N.J.; Lee, S.; Lee, E.W. Effects of HIPping on high-cycle fatigue properties of investment cast A356 aluminum alloys. *Mater. Sci. Eng. A* **2003**, *340*, 123–129. [CrossRef]
15. Wang, Q.; Apelian, D.; Lados, D. Fatigue behavior of A356/357 aluminum cast alloys. Part II—Effect of microstructural constituents. *J. Light Met.* **2001**, *1*, 85–97. [CrossRef]
16. Wang, Q.; Apelian, D.; Lados, D. Fatigue behavior of A356-T6 aluminum cast alloys. Part I. Effect of casting defects. *J. Light Met.* **2001**, *1*, 73–84. [CrossRef]
17. Beretta, S.; Romano, S. A comparison of fatigue strength sensitivity to defects for materials manufactured by AM or traditional processes. *Int. J. Fatigue* **2017**, *94*, 178–191. [CrossRef]
18. Brandl, E.; Heckenberger, U.; Holzinger, V.; Buchbinder, D. Additive manufactured AlSi10Mg samples using Selective Laser Melting (SLM): Microstructure, high cycle fatigue, and fracture behavior. *Mater. Des.* **2012**, *34*, 159–169. [CrossRef]
19. European Committee for Standardization (CEN). *Aluminium and Aluminium Alloys—Castings—Chemical Composition and Mechanical Properties*; CEN: Brussels, Belgium, 2010.
20. Gänser, H.-P.; Maierhofer, J.; Christiner, T. Statistical correction for reinserted runouts in fatigue testing. *Int. J. Fatigue* **2015**, *80*, 76–80. [CrossRef]
21. Ngnekou, J.N.D.; Henaff, G.; Nadot, Y.; Nicolai, J.; Ridosz, L. Fatigue resistance of selectively laser melted aluminum alloy under T6 heat treatment. *Procedia Eng.* **2018**, *213*, 79–88. [CrossRef]
22. Li, W.; Li, S.; Liu, J.; Zhang, A.; Zhou, Y.; Wei, Q.; Yan, C.; Shi, Y. Effect of heat treatment on AlSi10Mg alloy fabricated by selective laser melting: Microstructure evolution, mechanical properties and fracture mechanism. *Mater. Sci. Eng. A* **2016**, *663*, 116–125. [CrossRef]
23. Prashanth, K.G.; Scudino, S.; Klauss, H.J.; Surreddi, K.B.; Löber, L.; Wang, Z.; Chaubey, A.K.; Kühn, U.; Eckert, J. Microstructure and mechanical properties of Al-12Si produced by selective laser melting: Effect of heat treatment. *Mater. Sci. Eng. A* **2014**, *590*, 153–160. [CrossRef]
24. Takata, N.; Kodaira, H.; Sekizawa, K.; Suzuki, A.; Kobashi, M. Change in microstructure of selectively laser melted AlSi10Mg alloy with heat treatments. *Mater. Sci. Eng. A* **2017**, *704*, 218–228. [CrossRef]
25. Zhang, C.; Zhu, H.; Liao, H.; Cheng, Y.; Hu, Z.; Zeng, X. Effect of heat treatments on fatigue property of selective laser melting AlSi10Mg. *Int. J. Fatigue* **2018**, *116*, 513–522. [CrossRef]
26. Gall, K.; Yang, N.; Horstemeyer, M.; McDowell, D.L.; Fan, J. The debonding and fracture of Si particles during the fatigue of a cast Al-Si alloy. *Met. Mat. Trans. A* **1999**, *30*, 3079–3088. [CrossRef]
27. ASTM International. *Standard Practice for Statistical Analysis of Linear or Linearized Stress-Life (S-N) and Strain-Life (ε-N) Fatigue Data*; ASTM International: West Conshohocken, PA, USA, 2015.
28. Dengel, D. Die arc sin $\sqrt{P}$-Transformation—Ein einfaches Verfahren zur grafischen und rechnerischen Auswertung geplanter Wöhlerversuche. *Mater. Werkst.* **1975**, *6*, 253–261. [CrossRef]
29. EOS GmbH-Electro Optical Systems. Data Sheet: EOS Aluminium AlSi10Mg. 2014. Available online: https://lightway-3d.de/download/LIGHTWAY_EOS_Aluminium_AlSi10Mg_de_Datenblatt.pdf (accessed on 11 December 2018).
30. Masuo, H.; Tanaka, Y.; Morokoshi, S.; Yagura, H.; Uchida, T.; Yamamoto, Y.; Murakami, Y. Effects of Defects, Surface Roughness and HIP on Fatigue Strength of Ti-6Al-4V manufactured by Additive Manufacturing. *Procedia Struct. Integr.* **2017**, *7*, 19–26. [CrossRef]
31. Romano, S.; Beretta, S.; Brandão, A.; Gumpinger, J.; Ghidini, T. HCF resistance of AlSi10Mg produced by SLM in relation to the presence of defects. *Procedia Struct. Integr.* **2017**, *7*, 101–108. [CrossRef]

© 2019 by the authors. Licensee MDPI, Basel, Switzerland. This article is an open access article distributed under the terms and conditions of the Creative Commons Attribution (CC BY) license (http://creativecommons.org/licenses/by/4.0/).

Article

Dimensional Quality and Distortion Analysis of Thin-Walled Alloy Parts of AlSi10Mg Manufactured by Selective Laser Melting

Altaf Ahmed [1,*], Arfan Majeed [2,*], Zahid Atta [2] and Guozhu Jia [1]

1. Department of Management Science and Engineering, School of Economics and Management, Beihang University (BUAA), Beijing 100191, China; jiaguozhu@buaa.edu.cn
2. Key Laboratory of Contemporary Design and Integrated Manufacturing Technology, School of Mechanical Engineering, Northwestern Polytechnical University, Shaanxi 710072, China; zahid.atta@mail.nwpu.edu.cn
* Correspondence: altafahmed@buaa.edu.cn (A.A.); amajeed@mail.nwpu.edu.cn (A.M.)

Received: 21 May 2019; Accepted: 19 June 2019; Published: 21 June 2019

Abstract: The quality and reliability in additive manufacturing is an emerging area. To ensure process quality and reliability, the influence of all process parameters and conditions needs to be understood. The product quality and reliability characteristics, i.e., dimensional accuracy, precision, repeatability, and reproducibility are mostly affected by inherent and systematic manufacturing process variations. This paper presents research on dimensional quality and distortion analysis of AlSi10Mg thin-walled parts developed by a selective laser melting technique. The input process parameters were fixed, and the impact of inherent process variation on dimensional accuracy and precision was studied. The process stability and variability were examined under repeatability and reproducibility conditions. The sample length (horizontal dimension) results revealed a 0.05 mm maximum dimensional error, 0.0197 mm repeatability, and 0.0169 mm reproducibility. Similarly, in sample height (vertical dimension) results, 0.258 mm maximum dimensional error, 0.0237 mm repeatability, and 0.0863 mm reproducibility were observed. The effect of varying design thickness on thickness accuracy was analyzed, and regression analysis performed. The maximum 0.038 mm error and 0.018 mm standard deviation was observed for the 1 mm thickness sample, which significantly decreased for sample thickness ≥2 mm. The % error decreased exponentially with increasing sample thickness. The distortion analysis was performed to explore the effect of sample thickness on part distortion. The 0.5 mm thickness sample shows a very high distortion comparatively, and it is reduced significantly for >0.5 mm thickness samples. The study is further extended to examine the effect of solution heat treatment and artificial aging on the accuracy, precision, and distortion; however, it did not improve the results. Conclusively, the sample dimensions, i.e., length and height, have shown fluctuations due to inherent process characteristics under repeatability and reproducibility conditions. The ANOVA results revealed that sample length means are not statistically significantly different, whereas sample height means are significantly different. The horizontal dimensions in the xy-plane have better accuracy and precision compared to the vertical dimension in the z-axis. The accuracy and precision increased, whereas part distortion decreased with increasing thickness.

Keywords: dimensional quality analysis; repeatability and reproducibility; process variability; distortion analysis; selective laser melting

1. Introduction

Quality and reliability are major concerns in the state-of-the-art Industry 4.0 technologies including Additive Manufacturing (AM). AM technologies have gained more attention recently due to their ability to manufacture complex and fully functional geometries by sequential addition of material (layer-after-layer) beginning from 3D digital models. AM Research is in progress in multiple directions, and there are many quality related issues that are still challenging and need to be addressed [1]. Among AM technologies, selective laser melting (SLM) recently emerged as the widely used technique in aerospace, automotive and biomedical productions due to its ability to build complex parts and parts having open cell structures along with the minimum amount of material wastage [2–4]. Several parameters and conditions in the SLM process have uncertainties and varying effects on the final product. These process parameters and conditions are under investigation to achieve the desired level of quality and reliability [5–8].

The AlSi10Mg material, due to its hypoeutectic microstructure, is equivalent to A360 die-cast aluminum in additive manufacturing [5,6]. The thin-walled parts of AlSi10Mg due to their exceptional characteristics including low thermal expansion coefficient, less weight, stiffness, high specific strength, corrosion resistance, high thermal and electrical conductivities have found wide applications in aerospace, automobile, energy, electronics, and railway industries [7,9,10]. At present, conventional manufacturing techniques including extrusion, casting and forging are used to produce a significant portion of aluminum alloys part of complex geometries, like thin-walled and asymmetrical forms and internal flow capillaries, resulting in lengthy production hold-ups and higher expenditures [11]. Current industrial applications of AlSi10Mg need innovative production techniques. Selective laser melting, a type of powder bed fusion (PBF) is a favorable AM technique with benefits such as complex geometry design, production flexibility, as well as cost and time savings [12–14]. There are different sets of process parameters such as part placement, scanning direction, scanning strategy, inert gas flow velocity, laser power, part built-up direction, hatch spacing, scanning speed, powder bed temperature and layer thickness to control the microstructure and mechanical properties of AlSi10Mg manufactured thin-walled parts with selective laser melting (SLM) technique [9,15–18].

In AM processes, the dimensional variation among the computer aided designed part, and the actual built part is defined as geometrical accuracy. Due to the layer by layer building process, many factors affect the geometrical accuracy of the actual parts. The mechanical precision of the manufacturing setup, such as layer thickness, concentrated laser spot size, and scanner's position precision is amongst the factors affecting dimensional accuracy. The surface morphology that is described by numerous factors affects the geometrical accuracy as well. The factors mentioned above greatly depend upon the part positioning relative to the build direction [19]. Di W et al. [20] examined the geometrical characteristics of SLM built parts and concluded that the laser penetration, width of the laser beam, stair effect and powder adhesion play a key role in affecting the dimensional accuracy of different geometrical shapes produced by selective laser melting. Davidson et al. [21] focused upon SLM of duplex stainless steel powders and discovered that the geometrical precision varies with the direction. They found that the laser power and percent dimensional error are directly proportional and a geometrical error of 2–3% was reported on the average.

Calignano [22,23] investigated the dimensional accuracy of laser powder fusion using AlSi10Mg alloy and stated that the accuracy of parts produced is affected by the STL file, build direction, and process parameters. Thermal stress and the setting of process parameters have an impact on surface roughness and dimensional accuracy as well. Yap et al. [24] studied the effect of process parameters on the dimensional accuracy of parts developed on the PolyJet 3D printer by using three types of benchmarks and concluded that in order to develop thin walls successfully, the wall thickness should be greater than 0.4 mm. Raghunath and Pandey [25] in their study revealed that laser power and scan length are sources of deviation in the x-axis, laser power, and beam speed are sources of deviation in the y-axis, whereas, bed temperature, hatch spacing, and beam speed are sources of deviation in the z-axis. Han et al. [26] studied the effects of various process parameters upon geometrical accuracy and established that the precision can be enhanced by high scan speed that results in high density. Majeed et al. [27] investigated the dimensional and surface quality of parts-built by AM technique and optimized the process parameters. Zhu et al. [28] concluded that the thermal shrinkage would be higher for high laser power and low scan speed and smaller spacing. Furthermore, as compared to the x-y plane, the total shrinkage is significantly high in the z plane. Yu et al. [29] studied the influence of re-melting on surface roughness and porosity of AlSi10Mg parts developed by SLM and found a positive effect on both of these properties.

One of the main disadvantages of SLMed parts is residual stress that leads to part distortion. Distortion significantly affects the dimensional accuracy of a part and adversely hinders the efficient working of the built parts. Kruth et al. [30] concluded that residual stresses cause distortion that affects the geometrical accuracy of the physical parts. It happens due to locally focused energy distortion, resulting in high-temperature gradients, which happens while separating the built part from the substrate. Shiomi et al. [31] found that rapid cooling and heating produces a high-temperature gradient that further leads to the generation of thermal stress and hence, causes part distortion and cracks. Yasa et al. [32] and Beal et al. [33] investigated the effects of SLM process parameters and found that scan strategy has a significant role in cracks formation and distortion of built parts. Li et al. [34] focused on quick anticipation of distortion in SLMed parts by developing a Finite Element model. The experimental results also confirmed forecast distortions in different scan strategies. Shukzi Afazov et al. [35] forecast and compensated the distortion in large scale industrial parts by developing two models for distortion compensation. Keller et al. [36] attained quick simulation of part distortion by establishing a multi-scale modeling technique that implied an intrinsic strain obtained from a hatch model of several laser scans in selective laser melting.

The researchers in their studies have determined different optimized parameters for porosity, roughness, hardness, dimensions, etc., but in actual practice, even at the optimized setting, there is variation in these quality characteristics of developed parts. These variations can be determined by repeatability and reproducibility experimentation, and analysis. The part-quality characteristics, i.e., dimension accuracy, precision, and distortion, can vary in the different axis or directions or change with dimension. Furthermore, the surface treatment can improve some quality properties, i.e., hardness, porosity, etc., and it can also affect these characteristics. Therefore, exploration of these points is the main objective of this work.

2. Material and Experimental Method

AlSi10Mg power was used for the building of thin-walled specimens whose morphology is shown in Figure 1. Specimens were built on an SLM 280 HL system, which was equipped with two 400 W fiber lasers. The chemical composition of AlSi10Mg powder was 10.1 % Si, 0.30% Mg, 0.11 % Fe, <0.05% Ni and balance % aluminum. In this study, the processing parameters of 0.320 kW laser power, 0.90 m/s scanning speed, 25% overlap rate, 0.08 mm of hatch distance, 0.03 mm of layer thickness, vertical building direction, and 67° checkerboard scanning strategy were used [37].

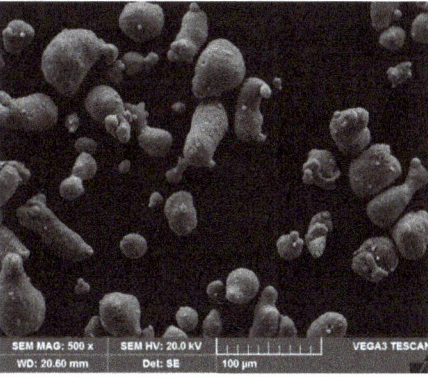

Figure 1. The morphology of AlSi10Mg powder particles.

The two dimensions, length (L), and height (H) of samples were fixed at 56 mm and 10.5 mm respectively, and the wall thickness of each sample was varied from 0.50 mm to 5.0 mm to make 12 combinations. Total 12 × 4 (4 Sets) samples were fabricated with a size of 56 mm × 10.5 mm × W_t; where W_t is wall thickness (i.e., 0.50, 0.80, 1.0, 1.20, 1.50, 1.80, 2.0, 2.50, 3.0, 3.5, 4.0, 5.0 mm). The third dimension thickness was systematically varied to study the effect of varying thickness on the dimensional quality and distortion.

The first three sets were fabricated in a single production run. The first set of 12 samples was used in As-Built (AB) condition for repeatability, reproducibility, and distortion analysis. The remaining two sets were analyzed after Solution Heat Treatment (SHT) and Artificial Aging (AA). The fourth set was fabricated at the same settings on the same system using the same material but at different intervals of time for the reproducibility study with the first set. The whole experimental scheme is presented in Figure 2. The repeatability and reproducibility were performed with the first and fourth set by using two dimensions, i.e., length (L) and height (H), which are fixed and produced at fixed input process parameters settings.

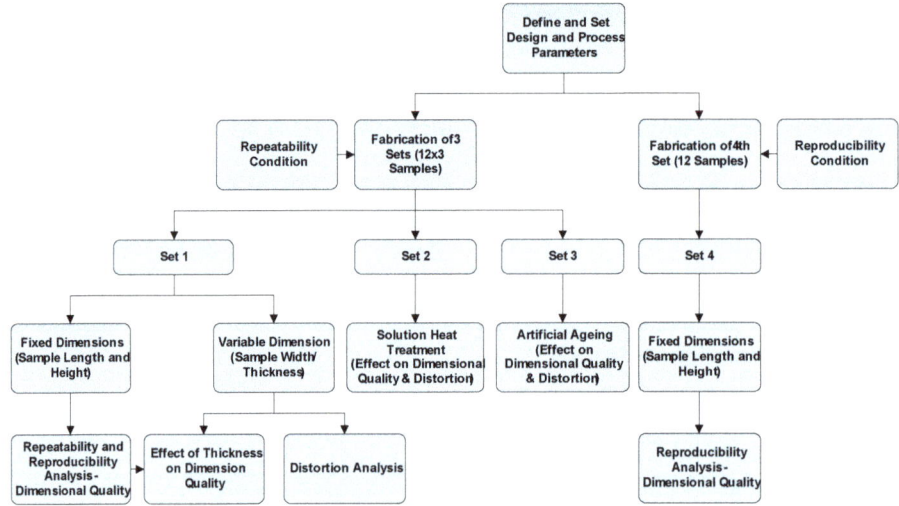

Figure 2. Experimental Scheme.

The scheme for sample build-up and reference directions is shown in Figure 3. The sample length and thickness are created in the *xy*-plane, horizontal direction. The sample height is created in the *z*-axis, vertical direction. The samples were separated from the substrate by using a wire cut electrical discharge machine. The developed samples and AM system are shown in Figure 4.

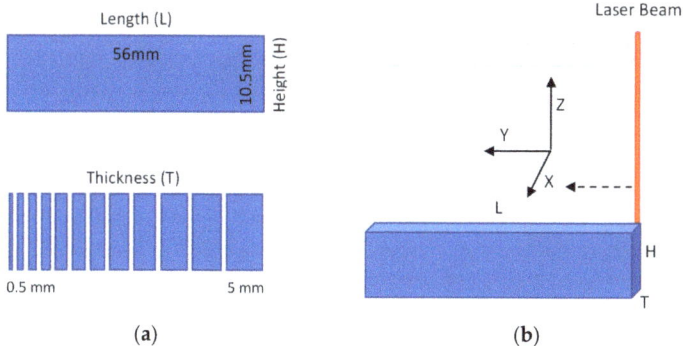

Figure 3. (a) Sample sizing and analyzed dimensions length (L), height (H), and thickness (T). (b) Sample build-up scheme.

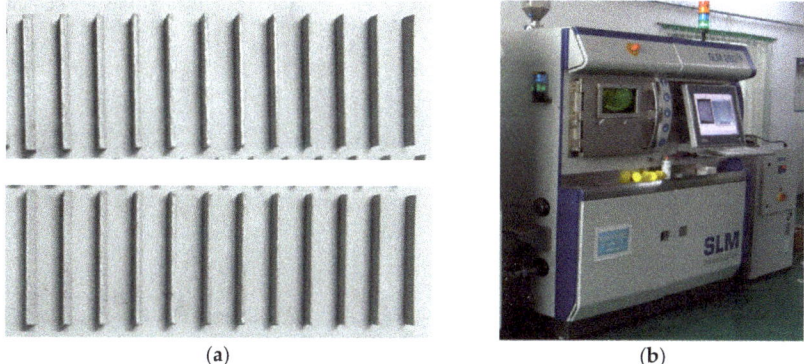

Figure 4. (a) Developed Samples (b) SLM 280HL System.

The length of each sample was measured three times and height five times; the width or thickness measured five times, and the average values were estimated. For distortion analysis, the sample was marked from one edge to another with ten positions 1 to 10 along the length of the sample. The distortion (displacement) values are measured at these marked positions to relate the measured values to the location of the sample.

The effect of heat treatments is also investigated on the thin-walled specimens by applying SHT and AA. Two sets were applied SHT at 530 °C and 540 °C for 2 h in the electric furnace, and the specimens were instantly exposed to water quenching at room temperature after SHT. AA was performed on 530 °C SHT set at 155 °C for 12 h in the drying oven, and further, the samples were quenched in the air to room temperature [38,39].

The powder morphology was tested with SEM Tescan VEGA3 LMU Scanning Electron Microscope system. The samples dimensional quality measurements were taken by using Mitutoyo vernier caliper, and their distortion was examined by using a dial indicator on a flatbed.

3. Results and Discussion

The results and discussion part is distributed into four sections. In the first section, we have fixed the input process parameters and determined the dimensional variations in 12 samples at as-built (AB) condition. The variation in the dimension of the parts depicts the manufacturing process variations at fixed conditions. The accuracy, precision, repeatability, and reproducibility are examined based on as-built samples considering two sides of the sample. In the second section, the variation in thickness accuracy with increasing sample thickness is presented. Further, correlation and regression analysis are studied. In the third section, distortion analysis is presented. The variation and correlation between distortion and sample thickness are discussed. Lastly, the effect of SHT and AA on sample quality characteristics, i.e., dimensional accuracy, precision, and distortion, are discussed. The analysis performed by using MINITAB 18, MATLAB 07R, and Origin Pro 9.

3.1. Dimensional Quality under Repeatability (Process Variability)

The repeatability is a condition in which parameters and conditions, i.e., machine, man, method and material, are fixed and the products are developed repeatedly, or values are taken in a short interval of time repeatedly, and it is represented numerically by the standard deviation. In our study, the design length (L) and height (H) of the samples are fixed at 56 mm and 10.5 mm, respectively. Twelve samples are developed at the fixed input process parameters under same conditions. The dimensional values of the length and height of as-built samples are measured and mentioned in Table 1. The length and height are the average value of three and five readings of each sample, respectively. As the inputs parameters and conditions are fixed, the estimated standard deviations in length and height data represent the repeatability of the production process.

Table 1. Measurement and ANOVA results of the samples (set 1) under repeatability condition.

	Sample Length (L)				Sample Height (H)		
Sample No	Design Length (mm)	Actual Mean Length (mm)	% Error	Sample No	Design Height (mm)	Actual Mean Height (mm)	% Error
1	56	55.977	0.042	1	10.5	10.438	0.590
2	56	56.013	0.024	2	10.5	10.530	0.286
3	56	55.990	0.018	3	10.5	10.544	0.419
4	56	55.973	0.048	4	10.5	10.664	1.562
5	56	55.997	0.006	5	10.5	10.564	0.610
6	56	55.973	0.048	6	10.5	10.528	0.267
7	56	55.983	0.030	7	10.5	10.504	0.038
8	56	55.990	0.018	8	10.5	10.422	0.743
9	56	56.000	0.000	9	10.5	10.472	0.267
10	56	56.007	0.012	10	10.5	10.464	0.343
11	56	56.007	0.012	11	10.5	10.470	0.286
12	56	56.010	0.018	12	10.5	10.426	0.705
Overall Mean Length (mm)			55.993	Overall Mean Height (mm)			10.502
Max. Error (mm)			0.027	Max Error (mm)			0.164
Repeatability σ_r (mm)			0.0197	Repeatability σ_r (mm)			0.0237
p-value			0.160	p-value			0.000
F-value			1.61	F-value			42.73

The accuracy and precision are estimated as sample error and standard deviation, respectively. The actual measured length observed between 55.977–56.013 mm, and the maximum error is 0.027 mm (0.048%). Similarly, the actual height observed between 10.422–10.664 mm, and the maximum error is 0.164 mm (1.562%).

Analysis of Variance (ANOVA) is performed to determine significant differences in the i) mean length and ii) mean height between the 12 samples of set 1 developed under fixed input parameter settings. Each sample has three values of length and five values of height. The *p*-value and *F*-value, mentioned in Table 1, shows the statistically significant difference between means and variation in means, respectively. The ANOVA performed at 95% confidence level by using the alpha level of 0.05. The normality of data checked, and normal probability plot of residuals indicated that residual data follow a normal distribution.

The repeatability estimated from ANOVA results, which is calculated by using the square root of mean squared error (MSE) value, also known as pooled standard deviation. The calculated repeatability (σ_r) for length and height is 0.0197 mm and 0.0237 mm, respectively.

Figure 5a shows the variation in the length of samples at repeatability condition. The interval on the bar represents the standard deviation, which is estimated based on three repeated values of each sample. The red line is the design or target length line. The results show a random distribution of values. It can be seen from the graph that the length of each sample is fluctuating and not consistent, which shows the degree of instability of the production process. Secondly, the target line falls within the standard deviation interval of most of the samples. The ANOVA results revealed that the length means of samples in set 1 are not statistically significantly different, which is indicated by *p*-value ($p = 0.160 > 0.05$).

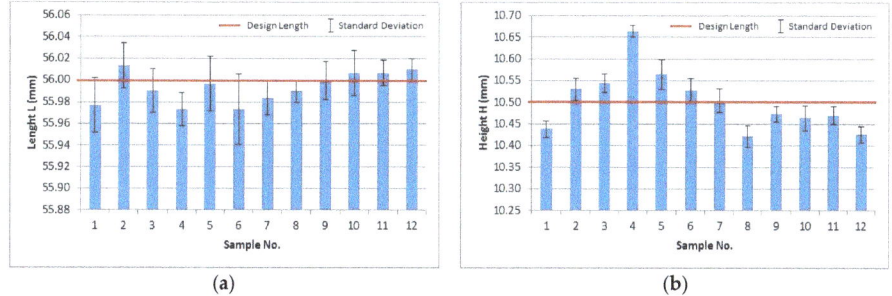

Figure 5. (a) Length and (b) height variation in developed samples with standard deviation.

Figure 5b shows the variation in the height of samples at repeatability condition. The standard deviation interval is calculated based on five repeated values of each sample. The height of each sample is inconsistent, which shows higher instability in the production process. The target line even falls within the standard deviation interval of only a few samples. The ANOVA results revealed that the height means of samples in set 1 are statistically significantly different, which is indicated by *p*-value ($p = 0.000 < 0.05$).

The height means values of some sample, i.e., 1, 4, 5, 8, and 12, are statistically significantly different as revealed from ANOVA results. The variation in the samples is due to the effect of solidification, random shrinkage behavior, and residual stresses. The layers can shrink non-uniformly due to low or high-temperature regions, and this non-uniformity shrinkage results in dimensional variations. The sample 4 and 5 have statistical significance and show a higher value than the target value. This may be due to laser heat, which penetrates more to bond unwanted powder particles. Further, it also can be attributed to the bed temperature variation as the part build at center or region of higher temperature, have larger dimension as compared to part build at the edge of the bed or the region of low temperature.

The results show variation or fluctuations in dimensional values and standard deviation, which are due to inherent random errors or effects of manufacturing process or system. It can be revealed from the results that the sample height is more inconsistent, have more error and standard deviation as compared to sample length. The sample length and width or thickness boundary is created as a result of the laser beam boundary in the *xy*-plane, as shown in Figure 3, whereas the sample height is in the *z*-axis, the direction in which the bed moves equal to one layer thickness and re-coater spread a new layer of powder. The sample dimension, which is created by a laser beam in the xy-plane, has more accuracy and precision as compared to the dimension created in the z-axis. This is because of internal stresses or shrinkage in xy-plane is lesser as compare to the z-axis, the vertical direction.

This shows that the variation of dimensional quality in different directions and the dimensions created in xy-plane will be more accurate and precise as compared to the dimension in the z-axis. This will help designers to achieve more accuracy in any specific part dimension which can be done by setting part build up a position in a direction that keeps the dimensions in the xy-plane that needs more accuracy and precision.

3.2. Dimensional Quality under Reproducibility (Process Variability)

The reproducibility is a condition in which one or more conditions are changed, i.e., machine, man, location, or time while keeping the same method and material. Two sets consisting of twelve samples in each set are developed at different time interval and production run. Table 2 shows the summarized results of both sets under reproducibility condition.

Table 2. Measurement and ANOVA results of set 1 and set 4 developed under reproducibility condition.

Parameter	Length (L)		Height (H)	
	Set 1	Set 4	Set 1	Set 4
Design Value (mm)	56	56	10.5	10.5
Mean Value (mm)	55.993	56.006	10.502	10.591
Max. Error in any sample (mm)	0.027 (0.048%)	0.050 (0.089%)	0.164 (1.564%)	0.258 (2.457%)
Reproducibility σ_R (mm)	0.0169		0.0863	
p-value	0.086 (>0.05)		0.019 (<0.05)	
F-value	3.23		6.39	

Analysis of Variance (ANOVA) is performed to determine significant differences in the i) mean length and ii) mean height between the set 1 and set 4 which are developed under fixed input parameters setting at different interval of time. Set 1 and set 4 considered as two groups having 12 values in each group. The p-value and F-value, mentioned in Table 2, shows the statistically significant difference between means and variation in means respectively. The ANOVA performed at 95% confidence level by using the alpha level of 0.05. The normality of data is checked, and a normal probability plot of residuals indicated that residual data follow a normal distribution.

The ANOVA results revealed that length means in set 1 and set 4 are not statistically significantly different which is indicated by p-value ($p = 0.086 > 0.05$) whereas the height means in set 1 and set 4 are statistically significantly different which is indicated by p-value ($p = 0.019 < 0.05$).

The reproducibility estimated from ANOVA results, which is calculated by using the square root of mean squared error (MSE) value, also known as pooled standard deviation. The calculated reproducibility (σ_R) is 0.0169 mm and 0.0863 mm for length and height, respectively.

The results revealed that the length and height show inconsistency and variability. The maximum dimensional error of 0.258 mm (2.457%) and a maximum standard deviation of 0.0863 mm observed under reproducibility condition. The height has less accuracy and precision as compared to the length and has shown the same trend as in repeatability condition.

3.3. Dimensional Quality with Variable Dimension

The dimensional quality is examined with the varying dimension. The sample design thickness is varied from 0.5 mm to 5 mm and, accuracy and precision are calculated from the actual thickness of the samples, as shown in Table 3. The results show that both % Error and the standard deviation are decreased with the increasing sample thickness. The maximum error of 0.038 mm is observed in the whole range.

Table 3. Measurement results of sample thickness.

Sample No	Design Thickness (mm)	Actual Thickness T (mm)			
		Mean Thickness (mm)	% Error	Max Error (mm)	Standard Deviation σ
1	0.5	0.488	2.40		0.0130
2	0.8	0.782	2.25		0.0045
3	1	0.962	3.80		0.0179
4	1.2	1.168	2.67		0.0084
5	1.5	1.494	0.40		0.0089
6	1.8	1.774	1.44	0.038	0.0055
7	2	1.994	0.30		0.0089
8	2.5	2.496	0.16		0.0089
9	3	3.008	0.27		0.0084
10	3.5	3.504	0.11		0.0055
11	4	4.006	0.15		0.0055
12	5	5.008	0.16		0.0045

The % Error value is random and higher in the region between 0.5 mm to 2 mm sample thickness. Whereas the % Error decrease and remain less than 0.30% in the region from 2 mm to 5 mm sample thickness, as shown in Figure 6. Similarly, the precision is higher with increasing the sample thickness. The results show that the dimensional quality will be better with increasing sample thickness, and it will be lower with decreasing thickness. This will be important for a product designer to consider these effects while designing the product, especially where a higher degree of accuracy and precision is required.

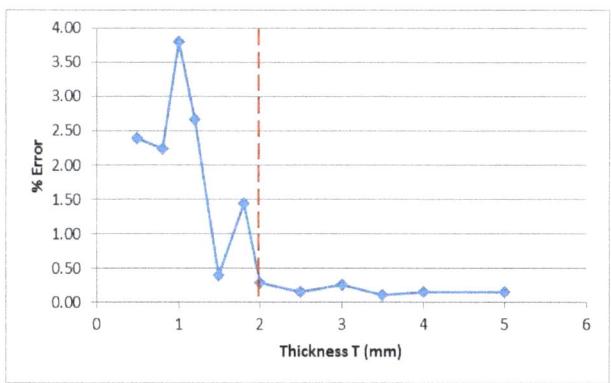

Figure 6. Experimental results show that % error in actual thickness decrease with increasing sample thickness. The % error reduces less than 0.3% for thickness greater than 2 mm.

The correlation and regression analysis are performed to determine the strength of the relationship between sample Thickness (T) and % Error. The correlation coefficient r is −0.73, which shows a negative relationship. The % Error decreased exponentially with the increasing thickness, which is presented by the regression model, as shown in Equation (1) and Figure 7. The R-squared value of the model is 0.6348 (63.48%). The p-value is 0.0006 (>0.05), which show the significance of the relationship.

$$\% \text{ Error} = 4.7792 \times \exp(-0.814255 \times T) \tag{1}$$

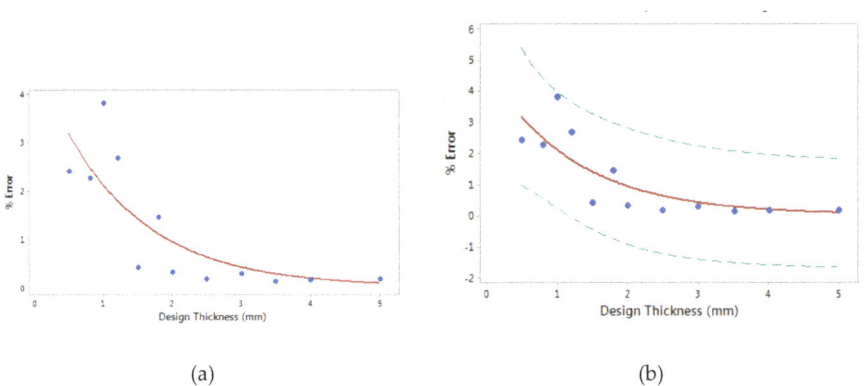

(a)   (b)

Figure 7. (a) Fitted line plot for a regression model. Sample thickness and % error in the thickness of developed samples decrease exponentially with increasing thickness. (b) Prediction plot showing the values falls within the 95% prediction interval. The central red line is fitted line, and the outer blue lines represent 95% prediction interval.

3.4. Distortion Analysis

The final quality of part depends on the material characteristics and production process parameters. The part deflection or distortion is a result of a combination of these factors. The residual stresses in a developed part cause the distortion. The distortion is measured by measuring the displacement using a dial indicator on a flatbed at ten points on each sample in as-built condition, and the results are shown in Table 4.

Table 4. Distortion measurement results of the thickness of samples.

Sample No	Design Thickness (mm)	Distortion (Displacement Measurement) mm										Mean	Std. Dev.	% Distorted
		1	2	3	4	5	6	7	8	9	10			
1	0.5	0.174	0.663	0.833	0.949	1.09	1.312	1.074	0.935	0.618	0.172	0.782	0.380	158.62
2	0.8	0.011	0.05	0.044	0.062	0.044	0.035	0.051	0.047	0.011	0.007	0.0362	0.020	4.54
3	1	0.055	0.191	0.082	0.125	0.116	0.199	0.135	0.192	0.154	0.079	0.1328	0.051	13.89
4	1.2	0.05	0.075	0.101	0.213	0.215	0.17	0.232	0.103	0.063	0.048	0.127	0.073	11.20
5	1.5	-0.023	0.022	0.034	0.04	0.087	0.005	0.01	0.001	-0.006	-0.021	0.0149	0.033	1.04
6	1.8	-0.003	0.017	0.002	0.014	-0.015	-0.009	-0.004	0.004	0.004	-0.009	0.0001	0.010	0.01
7	2	-0.008	0.011	0.019	0.02	0.033	0.015	0.006	0.025	0.033	0.108	0.0262	0.031	1.36
8	2.5	0.015	0.012	0.014	0.013	0.025	0.02	-0.001	0.004	0.025	0.019	0.0146	0.008	0.60
9	3	0.022	0.013	0.002	-0.008	-0.004	0.007	0.002	0.013	0.037	0.05	0.0134	0.018	0.45
10	3.5	0.043	-0.001	-0.013	-0.016	-0.034	-0.031	-0.031	-0.018	-0.012	0.001	-0.0112	0.023	0.33
11	4	0.029	0.017	0.007	0.006	0.004	0.003	0.025	-0.007	0	0.006	0.009	0.011	0.23
12	5	0.029	0.011	0.007	-0.001	-0.008	-0.01	-0.009	-0.017	-0.012	-0.019	-0.0029	0.015	0.06

The shrinkage value is subtracted from the measurement to get the actual distortion value. The positive and negative values indicate the side of deflection with reference to the central axis. The maximum mean distortion is 0.782 mm, and the maximum distortion at any single point is 1.312 mm, which is observed for 0.5 mm thickness. The maximum standard deviation of 0.038 mm is also observed for the 0.5 mm thickness sample.

Figure 8 shows the distortion variation or profile on the sample surface at 1 to 10 marked points. The distortion has higher values and variations in the region between 0.5–1.5 mm thicknesses. The 0.5 mm thickness sample has a maximum distortion and peak value at the middle location of the sample. The distortion considerably decreased after 0.5 mm sample thickness.

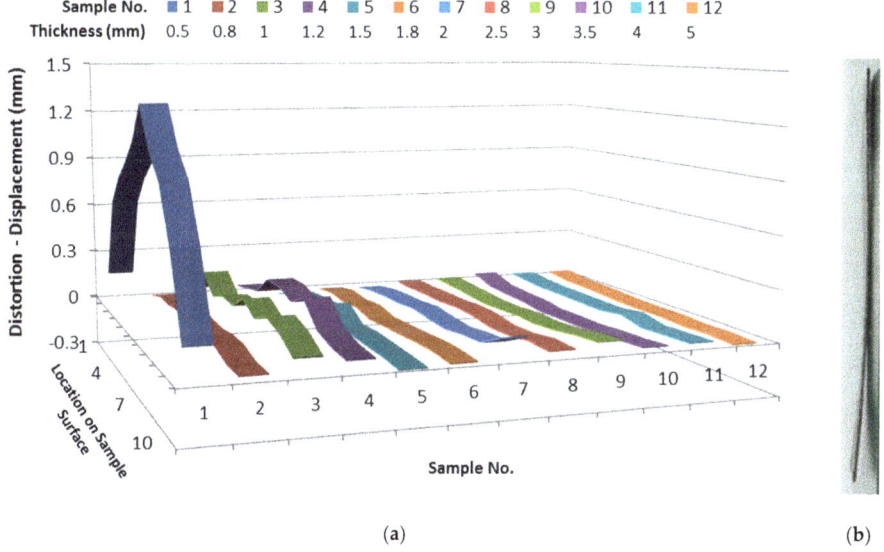

Figure 8. (a) The distortion profile is showing that it is decreasing with increasing sample thickness. (b) A sample showing distortion.

The differences in the distortion values are due to the residual stresses developed in the samples, that are the result of laser heat thermal cycling, i.e., heating and cooling during layer by layer development of samples. There is a temperature gradient between the bottom and each new upper layer. The thin samples are more prone to residual stresses, shrinkage, and bending as compared to thicker samples due to wall thickness, which cause higher distortion comparatively.

3.5. Effect of Heat Treatments (SHT and AA)

The samples are further analyzed to investigate the effect of SHT and AA on dimensional quality and distortion. Figures 9 and 10 show the results of % error and standard deviation in sample length and height under AB, SHT, and AA conditions. The result shows that SHT and AA have no clear effect on dimensional accuracy and precision. The results are random and do not depict any trend.

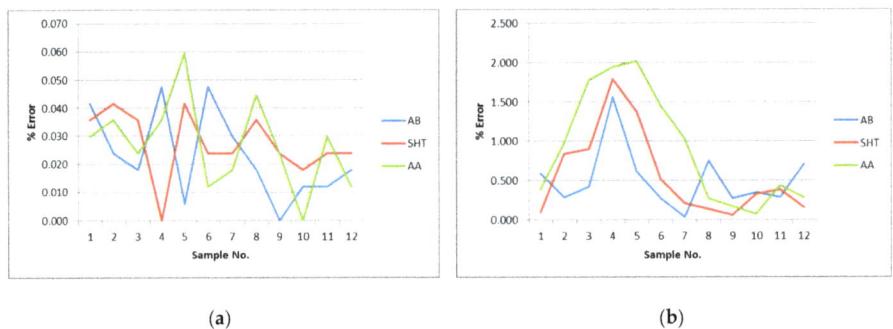

Figure 9. % Error in sample (**a**) length and (**b**) height comparison in As-Built (AB), Solution Heat Treatment (SHT), and Artificial Aging (AA) conditions.

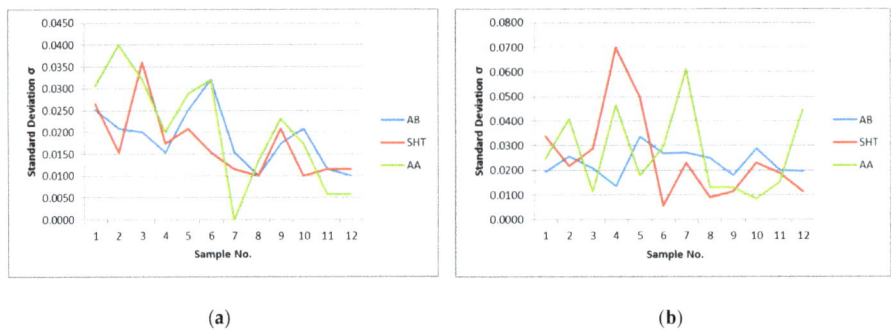

Figure 10. Standard Deviation σ in sample (**a**) length and (**b**) height comparison in AB, SHT, and AA conditions.

Figure 11 shows the results of distortion under AB, SHT, and AA aging conditions. The results are random and do not depict any beneficial effect of SHT and AA on distortion. Conclusively SHT and AA do not give any advantage in improving dimensional quality, i.e., accuracy and precision and reducing distortion.

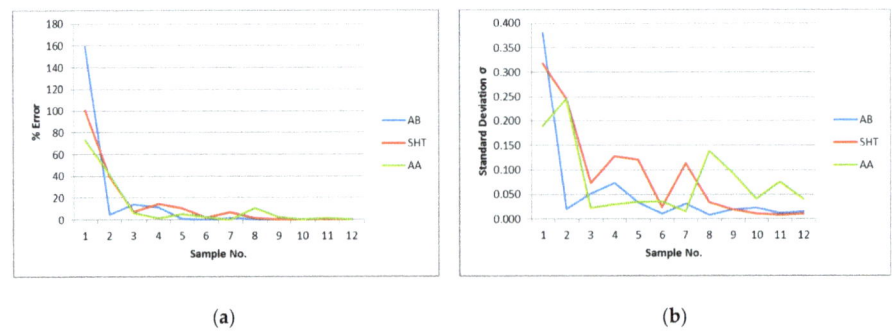

Figure 11. (**a**) Distortion % Error and (**b**) Distortion Standard Deviation Comparison under AB, SHT, and AA conditions.

4. Conclusions

In this paper, the dimensional quality, accuracy, and precision are investigated under repeatability and reproducibility conditions. The effect of increasing sample dimension, i.e., thickness, on the accuracy and precision, is studied followed by correlation and regression analysis. The distortion analysis is performed to examine the effect of SHT and AA for any improvement in dimensional quality and distortion. The following conclusive results are observed based on results and analysis;

- The manufacturing process has shown instability and random variations under repeatability condition, which is due to the inherent variability or random errors in the system.
- The dimensional quality results revealed that in sample length (horizontal dimension), 0.05 mm maximum dimensional error, 0.0197 mm repeatability (σ_r), and 0.0169 mm reproducibility (σ_R) observed. Similarly, in sample height (vertical dimension), 0.258 mm maximum error, 0.0237 mm repeatability (σ_r), and 0.0863 mm reproducibility (σ_R) observed.
- The ANOVA results revealed that length means (horizontal dimension) is not statistically significantly different under repeatability and reproducibility conditions. Whereas, the height means (vertical dimension) are statistically significantly different under repeatability and reproducibility conditions.
- The results show the variation of dimensional quality in horizontal and vertical directions. The dimensions created in xy-plane (horizontal direction) observed more accurate and precise as compared to the z-axis dimension (vertical direction).
- The dimensional error decreased with increasing sample thickness. The error reduces to less than 0.3% for thickness greater than 2 mm. The correlation analysis has revealed a negative correlation ($r = -0.73$) between % error and sample thickness. The regression model revealed an exponential decrease of %error with increasing thickness, $R_{sq} = 0.6348$ (63.48%), and p-value 0.0006 (<0.05), which shows the significance of the relationship.
- The sample distortion decreased with increasing sample thickness. The 0.5 mm thickness sample has shown very high distortion, whereas, the distortion reduced significantly for the 0.8–1.5 mm thickness samples.
- The solution heat treatment and artificial aging did not give any advantage in improving dimensional quality or reducing distortion in comparison with as-built condition results. It is not proven suitable for improvement purpose, but these HT conditions may improve other mechanical properties of parts like tensile strength, elongation, etc.

Author Contributions: Conceptualization, A.A., A.M. and G.J.; methodology, A.A., A.M.; samples fabrication and measurement, A.M. and Z.A.; validation and formal analysis, A.A. and A.M.; investigation, A.A., A.M. and Z.A.; data curation, A.A. and A.M.; writing—original draft preparation, A.A.; writing—review and editing, A.A., A.M. and Z.A.; visualization, A.A. and G.J.

Funding: This research was funded by the National Natural Science Foundation of China, grant number 51505423 and 51705428.

Acknowledgments: The authors would like to thank Yingfeng Zhang, Jingxiang Lv, and Tao Peng for their valuable guidance and support during this research.

Conflicts of Interest: The authors declare no conflict of interest.

References

1. Colosimo, B.M.; Huang, Q.; Dasgupta, T.; Tsung, F. Opportunities and challenges of quality engineering for additive manufacturing. *J. Qual. Technol.* **2018**, *50*, 233–252. [CrossRef]
2. Joshi, S.C.; Sheikh, A.A. 3D printing in aerospace and its long-term sustainability. *Virtual Phys. Prototyp.* **2015**, *10*, 175–185. [CrossRef]
3. Wong, K.K.; Ho, J.Y.; Leong, K.C.; Wong, T.N. Wong Fabrication of heat sinks by Selective Laser Melting for convective heat transfer applications. *Virtual Phys. Prototyp.* **2016**, *11*, 159–165. [CrossRef]

4. DebRoy, T.; Wei, H.L.; Zuback, J.S.; Mukherjee, T.; Elmer, J.W.; Milewski, J.O.; Beese, A.M.; Wilson-Heid, A.; De, A.; Zhang, W. Additive manufacturing of metallic components—Process, structure and properties. *Prog. Mater. Sci.* **2018**, *92*, 112–224. [CrossRef]
5. Vončina, M.; Kores, S.; Mrvar, P.; Medved, J. Effect of Ce on solidification and mechanical properties of A360 alloy. *J. Alloys Compd.* **2011**, *509*, 7349–7355. [CrossRef]
6. Li, X.P.; Wang, X.J.; Saunders, M.; Suvorova, A.; Zhang, L.C.; Liu, Y.J.; Fang, M.H.; Huang, Z.H.; Sercombe, T.B. A selective laser melting and solution heat treatment refined Al–12Si alloy with a controllable ultrafine eutectic microstructure and 25% tensile ductility. *Acta Mater.* **2015**, *95*, 74–82. [CrossRef]
7. Martin, J.H.; Yahata, B.D.; Hundley, J.M.; Mayer, J.A.; Schaedler, T.A.; Pollock, T.M. 3D printing of high-strength aluminium alloys. *Nature* **2017**, *549*, 365–369. [CrossRef]
8. Majeed, A.; Lv, J.; Zhang, Y.; Shamim, K.; Qureshi, M.E.; Muzamil, M.; Zafar, F.; Waqas, A. An investigation into the influence of processing parameters on the surface quality of AlSi10Mg parts by SLM process. In Proceedings of the 2019 16th International Bhurban Conference on Applied Sciences and Technology (IBCAST-2019), Islamabad, Pakistan, 8–12 January 2019.
9. Kempen, K.; Thijs, L.; Van Humbeeck, J.; Kruth, J.P. Processing AlSi10Mg by selective laser melting: Parameter optimisation and material characterisation. *Mater. Sci. Technol.* **2015**, *31*, 917–923. [CrossRef]
10. Jung, J.G.; Ahn, T.Y.; Cho, Y.H.; Kim, S.H.; Lee, J.M. Synergistic effect of ultrasonic melt treatment and fast cooling on the refinement of primary Si in a hypereutectic Al–Si alloy. *Acta Mater.* **2018**, *148*, 31–40. [CrossRef]
11. Cáceres, C.H.; Davidson, C.J.; Griffiths, J.R. The deformation and fracture behaviour of an Al-Si-Mg casting alloy. *Mater. Sci. Eng. A* **1995**, *197*, 171–179. [CrossRef]
12. Gu, D.D.; Meiners, W.; Wissenbach, K.; Poprawe, R. Laser additive manufacturing of metallic components: Materials, processes and mechanisms. *Int. Mater. Rev.* **2012**, *57*, 133–164. [CrossRef]
13. Herzog, D.; Seyda, V.; Wycisk, E.; Emmelmann, C. Additive manufacturing of metals. *Acta Mater.* **2016**, *117*, 371–392. [CrossRef]
14. Olakanmi, E.O.; Cochrane, R.F.; Dalgarno, K.W. A review on selective laser sintering/melting (SLS/SLM) of aluminium alloy powders: Processing, microstructure, and properties. *Prog. Mater. Sci.* **2015**, *74*, 401–477. [CrossRef]
15. Anwar, A.B.; Pham, Q.C. Selective laser melting of AlSi10Mg: Effects of scan direction, part placement and inert gas flow velocity on tensile. *J. Mater. Process. Technol.* **2017**, *240*, 388–396. [CrossRef]
16. Trevisan, F.; Calignano, F.; Lorusso, M.; Pakkanen, J.; Aversa, A.; Ambrosio, E.; Lombardi, M.; Fino, P.; Manfredi, D. On the Selective Laser Melting (SLM) of the AlSi10Mg Alloy: Process, Microstructure, and Mechanical Properties. *Materials* **2017**, *10*, 76. [CrossRef]
17. Krishnan, M.; Atzeni, E.; Canali, R.; Calignano, F.; Manfredi, D.; Ambrosio, E.P.; Iuliano, L. On the effect of process parameters on properties of AlSi10Mg parts produced by DMLS. *Rapid Prototyp. J.* **2014**, *20*, 449–458. [CrossRef]
18. Read, N.; Wang, W.; Essa, K.; Attallah, M.M. Selective laser melting of AlSi10Mg alloy: Process optimisation and mechanical properties development. *Mater. Des.* **2015**, *65*, 417–424. [CrossRef]
19. Strano, G.; Hao, L.; Everson, R.M.; Evans, K.E. Surface roughness analysis, modelling and prediction in selective laser melting. *J. Mater. Process. Technol.* **2013**, *213*, 589–597. [CrossRef]
20. Wang, D.; Wu, S.; Bai, Y.; Lin, H.; Yang, Y.; Song, C. Characteristics of typical geometrical features shaped by selective laser melting. *J. Laser Appl.* **2017**, *29*, 022007. [CrossRef]
21. Davidson, K.; Singamneni, S. Selective Laser Melting of Duplex Stainless Steel Powders: An Investigation. *Mater. Manuf. Process.* **2016**, *31*, 1543–1555. [CrossRef]
22. Calignano, F.; Lorusso, M.; Pakkanen, J.; Trevisan, F.; Ambrosio, E.P.; Manfredi, D.; Fino, P. Investigation of accuracy and dimensional limits of part produced in aluminum alloy by selective laser melting. *Int. J. Adv. Manuf. Technol.* **2017**, *88*, 451–458. [CrossRef]
23. Calignano, F. Investigation of the accuracy and roughness in the laser powder bed fusion process. *Virtual Phys. Prototyp.* **2018**, *13*, 97–104. [CrossRef]
24. Yap, Y.L.; Wang, C.; Sing, S.L.; Dikshit, V.; Yeong, W.Y.; Wei, J. Material jetting additive manufacturing: An experimental study using designed metrological benchmarks. *Precis. Eng.* **2017**, *50*, 275–285. [CrossRef]
25. Raghunath, N.; Pandey, P.M. Improving accuracy through shrinkage modelling by using Taguchi method in selective laser sintering. *Int. J. Mach. Tools Manuf.* **2007**, *47*, 985–995. [CrossRef]

26. Han, X.; Zhu, H.; Nie, X.; Wang, G.; Zeng, X. Investigation on Selective Laser Melting AlSi10Mg Cellular Lattice Strut: Molten Pool Morphology, Surface Roughness and Dimensional Accuracy. *Materials* **2018**, *11*, 392. [CrossRef] [PubMed]
27. Majeed, A.; Lv, J.; Peng, T. A framework for big data driven process analysis and optimization for additive manufacturing. *Rapid Prototyp. J.* **2019**, *25*, 308–321. [CrossRef]
28. Zhu, H.H.; Lu, L.; Fuh, J.Y.H. Study on shrinkage behaviour of direct laser sintering metallic powder. *Proc. Inst. Mech. Eng. Part B J. Eng. Manuf.* **2006**, *220*, 183–190. [CrossRef]
29. Yu, W.; Sing, S.L.; Chua, C.K.; Tian, X. Influence of re-melting on surface roughness and porosity of AlSi10Mg parts fabricated by selective laser melting. *J. Alloys Compd.* **2019**, *792*, 574–581. [CrossRef]
30. Kruth, J.-P.; Badrossamay, M.; Yasa, E.; Deckers, J.; Thijs, L.; Van Humbeeck, J. Part and Material Properties in Selective Laser Melting of Metals. In Proceedings of the 16th International Symposium on Electromachining (ISEM–XVI), Shanghai, China, 19–23 April 2010.
31. Shiomi, M.; Osakada, K.; Nakamura, K.; Yamashita, T.; Abe, F. Residual Stress within Metallic Model Made by Selective Laser Melting Process. *CIRP Ann.* **2004**, *53*, 195–198. [CrossRef]
32. Yasa, E.; Deckers, J.; Craeghs, T.; Badrossamay, M.; Kruth, J.P. Investigation on occurrence of elevated edges in selective laser melting. In Proceedings of the 20th Annual International Solid Freeform Fabrication Symposium, SFF 2009, Austin, TX, USA, 3–5 August 2009; pp. 673–685.
33. Beal, V.E.; Erasenthiran, P.; Hopkinson, N.; Dickens, P.; Ahrens, C.H. Scanning strategies and spacing effect on laser fusion of H13 tool steel powder using high power Nd: YAG pulsed laser. *Int. J. Prod. Res.* **2007**, *46*, 217–232. [CrossRef]
34. Li, C.; Fu, C.H.; Guo, Y.B.; Fang, F.Z. Fast Prediction and Validation of Part Distortion in Selective Laser Melting. *Procedia Manuf.* **2015**, *1*, 355–365. [CrossRef]
35. Afazov, S.; Denmark, W.A.; Toralles, B.L.; Holloway, A.; Yaghi, A. Distortion Prediction and Compensation in Selective Laser Melting. *Addit. Manuf.* **2017**, *17*, 15–22. [CrossRef]
36. Keller, N.; Ploshikhin, V. New method for fast predictions of residual stress and distortion of AM parts. In Proceedings of the Solid Freeform Fabrication Symposium, Austin, TX, USA, 4–6 August 2014; Volume 25, pp. 4–6.
37. Majeed, A.; Zhang, Y.; Lv, J.; Peng, T.; Waqar, S.; Atta, Z. Study the effect of heat treatment on the relative density of SLM built parts of AlSi10Mg alloy. In Proceedings of the 48th International Conference on Computers and Industrial Engineering (CIE 2018), Auckland, New Zealand, 2–5 December 2018.
38. Aboulkhair, N.T.; Tuck, C.; Ashcroft, I.; Maskery, I.; Everitt, N.M. On the Precipitation Hardening of Selective Laser Melted AlSi10Mg. *Metall. Mater. Trans. A* **2015**, *46*, 3337–3341. [CrossRef]
39. Li, W.; Li, S.; Liu, J.; Zhang, A.; Zhou, Y.; Wei, Q.; Yan, C.; Shi, Y. Effect of heat treatment on AlSi10Mg alloy fabricated by selective laser melting: Microstructure evolution, mechanical properties and fracture mechanism. *Mater. Sci. Eng. A* **2016**, *663*, 116–125. [CrossRef]

© 2019 by the authors. Licensee MDPI, Basel, Switzerland. This article is an open access article distributed under the terms and conditions of the Creative Commons Attribution (CC BY) license (http://creativecommons.org/licenses/by/4.0/).

Article

Intra- and Inter-Repeatability of Profile Deviations of an AlSi10Mg Tooling Component Manufactured by Laser Powder Bed Fusion

Floriane Zongo, Antoine Tahan *, Ali Aidibe and Vladimir Brailovski

Department of Mechanical Engineering, École de Technologie Supérieure (ÉTS),
Montreal, QC H3C 1K3, Canada; teega-wende-floriane.zongo.1@etsmtl.net (F.Z.); ali.aidibe.1@etsmtl.net (A.A.); vladimir.brailovski@etsmtl.ca (V.B.)
* Correspondence: antoine.tahan@etsmtl.ca; Tel.: +1-514-396-8687

Received: 13 July 2018; Accepted: 15 August 2018; Published: 21 August 2018

Abstract: Laser powder bed fusion (LPBF) is one of the most potent additive manufacturing (AM) processes. Metallic LPBF is gaining popularity, but one of the obstacles facing its larger industrial use is the limited knowledge of its dimensional and geometrical performances. This paper presents a metrological investigation of the geometrical and dimensional deviations of a selected LPBF-manufactured component, according to the ASME Y14.5-2009 standard. This approach allows for an estimation of both the process capability, as per ISO 22514-4 standard, and the correlations between the part location in the manufacturing chamber and the profile deviations. Forty-nine parts, which are representative of a typical aerospace tooling component (30 mm in diameter and 27.2 mm in height) were manufactured from AlSi10Mg powder using an EOSINT M280 printer and subjected to a stress relief annealing at 300 °C for two hours. This manufacturing procedure was repeated three times. A complete statistical analysis was carried out and the results of the investigation show that LPBF performances for all geometrical variations of 147 identical parts fall within a range of 230 μm at a 99.73% level.

Keywords: additive manufacturing; laser powder bed fusion; selective laser melting; metrology; inter-repeatability; intra-repeatability; geometrical dimensioning and tolerancing (GD and T); process capability

1. Introduction

Additive manufacturing (AM) technologies produce 3D engineered parts from nominal CAD files in an additive manner, generally layer by layer. The term "additive" is used to highlight the fact that these technologies do not require conventional tooling to build components and that the shape is created by adding, rather than removing or deforming, material. The material can be polymer, metal, composite, ceramic, concrete, or even human cells. Many AM processes have been developed and are commercially available, including stereolithography (SL), fused deposition modeling (FDM), three-dimensional printing (3DP), powder bed fusion (PBF), direct metal deposition (DED), and sheet lamination (SL). The PBF technologies include two variants depending on the nature of the heat source: the electron beam powder bed fusion (EBPBF) and the laser powder bed fusion (LPBF). Their general principles are described on ISO/ASTM52901-16 [1]. The processes terminologies used are from ISO/ASTM 52900:2015 [2] standard terminology for AM.

Wohler's report stated that 13,058 AM machines were sold in 2016 [3]. The use of these processes is expanding and can be explained by the benefits they provide: free complexity and easy customization, as well as the reduced setup time, delivery time, and tooling cost. LPBF is one of the most potent metallic AM technologies. However, the laser power, temperature field heterogeneity, and other phenomena inherent to the process generate residual stresses responsible for distortions of the produced parts [4]. Geometrical and dimensional deviations (GD and T) in LPBF parts are among the main concerns as far as it concerns facing wider industrial application of this technology. There is a need to study the process and improve part precision, which has been criticized by many researchers.

Wang et al. [5] studied the correlations between shrinkage, laser beam offset, and the weight of LPBF parts. After statistical analysis, sampling theory and three calculation methods, the conclusion was that the shrinkage remains nearly unchanged irrespective of the weight of AM parts. However, the beam offset increases with part weight. One of the first shrinkage calibrators for metallic AM was also proposed. Zhu et al. [6] studied the shrinkage of direct laser sintered metallic powder parts. Two types of shrinkage, thermal and sintering shrinkage, were isolated and quantified. Thermal shrinkage results from cyclic heating, while sintering shrinkage is caused by densification and is a type of elastic compressive shortening. The conclusion was that the higher the laser power and the smaller the scan speed and spacing, the higher the thermal shrinkage. Additionally, the total shrinkage in the Z plane is significantly higher than in the X-Y planes.

Raghunath and Pandey [7] identified the sources of deviation for each build axis using the Analysis of Variance (ANOVA) technique. Laser power and scan length were identified as the primary sources of deviations in the X-axis, laser power and beam speed in the Y-axis, and part bed temperature, hatch spacing and beam speed in the Z-axis. Islam and Shacks [8] investigated the influence of build parameters on the dimensional errors of 60 selective laser sintered polyamide parts. Senthilkumaran et al. [9] developed a model for shrinkage compensation in LPBF which operates in each layer. Galovskyi et al. [10] tested some work pieces for LPBF.

Detailed investigations of AM part geometrical deviations have been carried out in [11–23]. Fahad and Hopkinson [24] proposed a benchmark to evaluate and compare the accuracy and repeatability of the AM processes. This benchmark has three repetitions of features with standard geometries. With the intention of testing the LPBF process, Teeter et al. [25] conducted a metrological study about deviations appearing according to part location in the manufacturing chamber. After printing five pattern repetitions on a plate (the inspection was performed using an Olympus microscope with a resolution of ±0.5 μm), there was no difference between the pattern profile deviations. Ferrar et al. [26] investigated the gas flow effect on SLS repeatability and performance. In their study, variations in gas flows have been shown to affect both the value, the density and the compression strength range of the samples tested. Aidibe et al. [27] investigated the repeatability of the LPBF technology with five Ti-6Al-4V parts. The conclusion was that the LPBF process can provide acceptable metrological performances in terms of repeatability, overall deviations and geometric/dimensional errors, comparable to turning. Rebaioli and Fassi [28] identified some benchmark artefacts designed to evaluate the geometrical performance of the AM processes and their design guidelines. Sing et al. [29] investigated the effect of LPBF processing parameters on the dimensional accuracy and mechanical properties of cellular lattice structure using a statistical modeling. The conclusion was that the strut dimensions of LPBF fabricated lattice structures are most sensitive to laser power, as compared to layer thickness and scanning speed. Calignano [30] investigated the accuracy and surface roughness of parts manufactured by LPBF in the AlSi10Mg powder. The conclusion was that the STL file, build orientation, and process parameters affects the accuracy.

Globally, researchers have focused more on feasibility rather than on capability studies, the former revealing process limitations in printing some specific geometric features, while the latter provides an estimation of the probabilistic behavior of some metrological characteristics of the part produced by this process. Since the latter aspect represents a main goal of this study, this paper quantifies the LPBF process intra and inter repeatability, and capability with AlSi10Mg powders. The paper is organized as follows: Section 2 describes the experimental procedure. The results are presented in Section 3 and discussed in Section 4. Finally, a summary is provided and future works are presented in Section 5.

2. Experimental Protocol

The first goal of the experimental procedure is to identify and quantify the variations in the geometrical deviations of a selected part as a function of its location in the LPBF manufacturing chamber. Then, this experiment is intended to provide an answer to the hypothesis of a repeatable pattern of such deviations.

To this end, 49 identical AlSi10Mg parts equally distributed on a build plate (Figure 1) were printed three (3) times in the same LPBF system using the same process and post-process parameters, and analyzed by the same operator using the same equipment. The printed part is a typical aerospace tooling component, 30 mm in diameter and 27.2 mm in height. This part was chosen because it is an industrial tooling component used in jig construction, it is a kind of case study for industries interested in manufacturing by LPBF. Secondly, it is a topologically-optimized part. Finally, this part allows us to have an adequate sample size (49 parts/plate) for our study. Since we are concerned by GD and T variations as a function of part location in the fabrication chamber, an interesting element of this study is the number of repetitions which is 49 times three (49 × 3). This means that information from 49 different emplacements on the plate quantifies the variations occurring at the same place three times.

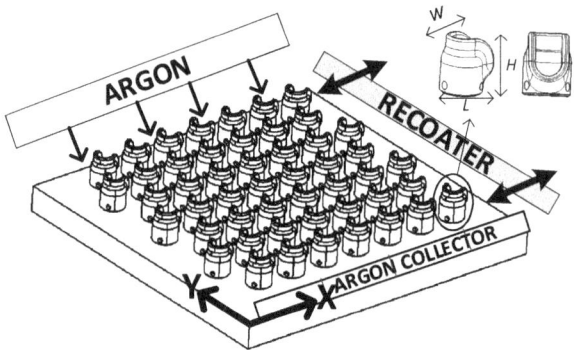

Figure 1. Parts disposition in the chamber for each build (EOS M 280).

In most cases, to reduce the risk of distortions caused by thermal gradients, while firmly attaching the part to the build plate during printing, the part needs to be built with support structures. In this study, specialized software Magics v.17.02 (Materialise, Leuven, Belgium) was used to generate support structures. The assembly was then loaded in the process software (PSW.3.4), where it was duplicated 49 times. The process parameters set, called AlSi10Mg_Speed 1.0 and recommended by the manufacturer EOS (Krailling, Germany) for an AlSi10Mg alloy, was used, with 30 μm-thick layers (Figure 2a). After printing, the build plate was stress relieved at 300 °C for two hours with no visible effect on the outer surface of the parts (Figure 2b).

Next, the point cloud of printed parts was obtained by means of a Metris LC50 laser scan mounted on a Mitutoyo Coordinate Measuring Machine (CMM) (accuracy $\leq$7 μm at the 95% level), Figure 2c. Before each scan, the devices were calibrated using a master sphere and the data collection was performed on nine (9) angles to maximize the information collection on inner surfaces. A real-time visualization was possible with the Focus Inspector specialized software. A thin layer of talcum powder was used to reduce part surface reflection. In doing so, the potential point cloud density was increased to ensure the best measurement. The point clouds was then assembled (from the nine angles) and cleaned. The parts were scanned before and after being cut off the plate. The best-fit technique was then carried out using PolyWorks® v.16 (Innovmetric Metrological Software, Quebec, QC, Canada). The data were then loaded into a Matlab® 2017b (software of MathWorks, Natick, MA, USA), using a code to extract the deviation at each point. Minitab® v.17 (a statistical software of Minitab Inc., State College, PA, USA) was used for the graphics and statistical studies (Figure 2d).

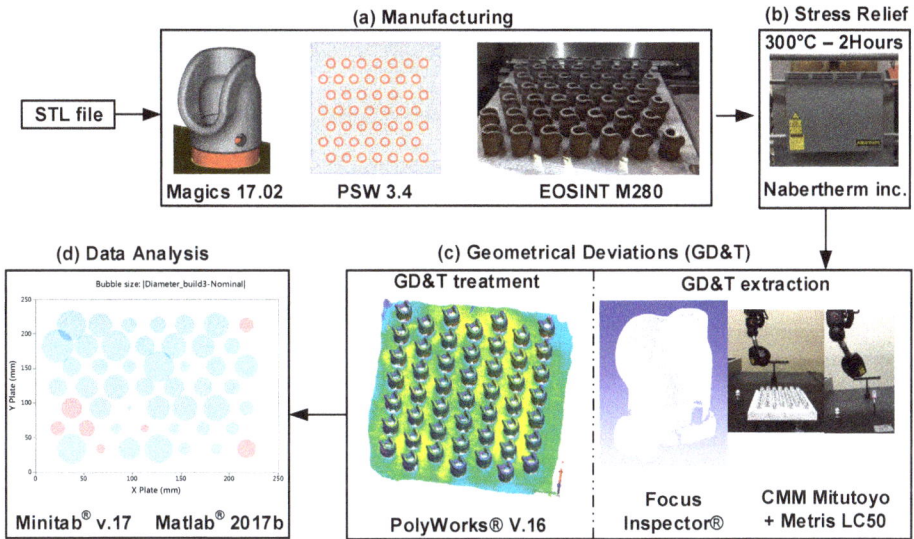

Figure 2. Experimental protocol: (**a**) manufacturing sequence, (**b**) stress relief heat treatment, (**c**) geometrical deviation measurements, and (**d**) data analysis.

Four types of analysis were performed based on ASME Y14.5 (2009): Intra-build variation study (Analysis 1), inter-build variation study (Analysis 2), and a capability study according to ISO 22514-4 (Analysis 3).

2.1. Intra-Build Variations Study

The intra-build variation study (Analysis 1) consisted of measuring the profile deviations (without a frame of reference) between the digitized parts (SCAN) and the nominal part (CAD). The digitization provided an average of 400,000 points for each part. The overall 3D profile deviations were extracted using the IMInspect module of PolyWorks® v.16 for each part, and represented by their nonparametric medians. In the first part of this intra-build variation study (Analysis 1a), visualizing the repartition of the profile deviations in the manufacturing chamber was the main interest. The second object of interest was the deviations of the external diameter of the parts at a height of $z = 1.2$ mm (Analysis 1b). This plan $z = 1.2$ mm has been chosen because it is the mid-value between the chamfer and the holes in the cylindrical feature of the part. For each of the 147 parts, the absolute difference between the measured diameter (using best fit criteria) and the nominal diameter (⌀19.05 mm) was extracted using the IMInspect module of PolyWorks® v.16 and plotted using Minitab® v.17. The Analysis 1c consisted of a correlation study of the two previous variables, the overall 3D profile deviation and the external diameter at a height of $z = 1.2$ mm. This analysis was carried out using a regression equation, which is an algebraic representation of the regression line used to describe the relationship between the response and predictor variables. In our case, the measured diameter was used as a predictor variable, while the overall 3D profile deviation represented by its median was considered as a response variable. Minitab v.17 linear regression analysis was used to obtain the equations for the three builds. Finally, a basic statistical study was also conducted with the overall 3D profile deviations and the external diameter at a height of $z = 1.2$ mm (Analysis 1d).

2.2. Inter-Build Variations Study

In order to quantify the inter-build variations (Analysis 2), which is the variation behavior among three builds, two statistical analyses were performed: the Kolmogorov–Smirnov (KS) test (Analysis 2a) and the inter-repeatability quantification (Analysis 2b). A visual comparison was also carried out using the best-fit technique with PolyWorks® v.16. The KS test and visual comparison were performed using the data acquired before cutting the parts off the plate for Build #2 and Build #3 (Build #1 data before cutting the parts were not available). The KS test is a nonparametric goodness-of-fit test that compares cumulative distribution functions (CDF). It is explained below in Equations (1)–(3). In this case, the KS test was used to compare the CDF of the 3D profile deviation of Build #2 and Build #3 acquired before the part removal.

Given n data points $x_1, x_2, \ldots, x_n$ of the build #j, the empirical CDF is defined as:

$$F_{j,\,n_j}(t) = \frac{1}{n_j} \sum_{i=1}^{n_j} 1_{x_i \leq t} \tag{1}$$

where 1_{x_i} is the indicator of event x_i, n_j is the data size from build #j, and $F_{j,n_j}(t)$ is its corresponding empirical CDF. The KS test between Build #2 and Build #3 is based on the maximum distance between two curves:

$$KS_{n_2,n_3} = \sup_t \left| F_{2,n_2}(t) - F_{3,n_3}(t) \right| \tag{2}$$

The null hypothesis H_0 is F_{2,n_2} and F_{3,n_3} have identical CDF behavior. H_0 is rejected at a significance level $1 - \alpha$ if:

$$KS_{n_2,n_3} > c(1 - \alpha) \sqrt{(n_2 + n_3)/n_2 n_3} \tag{3}$$

where $c(1 - \alpha)$ is the inverse of the KS distribution at level $1 - \alpha$. The p-value is used as criteria for acceptance/rejection of the KS test. α is the type I error [31]. The significance level is this study is 95%. This significance level was chosen because he usually used in metrological analyses. If the p-value is lower than the significance level $\alpha = 0.05$, then the null hypothesis H_0 is rejected.

Analysis 2b is an inter-repeatability statistical study carried out using CDF of the 3D profile deviation of each part as shown in Equations (4)–(6). Nine (9) different locations were selected (to be specified below) to uniformly cover the build space. The inter-variation study was performed for each position at a 95% level:

$$PV = \pm 1.96 \sigma_{PV} \tag{3}$$

$$\sigma_{PV} = K_3 R \tag{4}$$

$$R = \max(x_i) - \min(x_i) \tag{5}$$

With x_i is the capabilities (as described in Equation (7)) of the profile deviation at location i for Build #j (1, 2, and 3), R is the range of the three parts, σ_{PV} is the standard deviation, and PV is the part variation. For this case, $K_3 = 0.5231$ [27].

2.3. Capability Study

According to the ISO 22514-4, the process capability is a statistical estimate of the outcome of a characteristic of a process which has been demonstrated to be in a state of statistical control (stable) and which describes the process ability to fulfill the requirements of a given characteristic. By definition, process capability is the interval between $L_1 = 0.135\%$ and $L_2 = 99.865\%$ of the individual values' distributions; in other words, the interval containing 99.73% of the data (Figure 3).

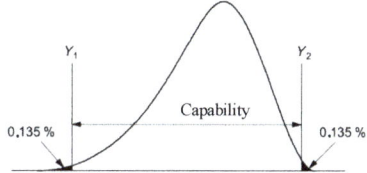

Figure 3. Capability interval in conformity with ISO 22514-4.

The capability study (Analysis 3) was performed using the non-parametric quantiles $L_{x\%}$ of the parts' profile deviations (Analysis 3a). The capability was obtained by:

$$\text{Capability} = L_{99.865\%} - L_{0.135\%} \tag{6}$$

Monte Carlo Simulation (MCS) [32] of the 3D profile deviation behavior was also carried out using Matlab® 2017b. For each part, the overall deviations were fitted to a normal distribution at a 95% confidence level. The MCS was then performed on the 147 normal distribution parameters, and the overall capability was extracted (Analysis 3b).

3. Results

The GD and T analysis was based on ASME Y14.5 (2009) and provides the following information: (1) Nonparametric intra-build variations study; (2) inter-build variations study, including goodness-to-fit test and; (3) capability study according to ISO 22514-4.

3.1. Intra-Build Variations

In the first study, each build is analyzed independently. This intra-build variation values are related to the location of each of the 49 parts uniformly distributed on the build plate and covering it entirely. In Analysis 1a, different colors are allocated to the deviation map shown in Figure 4 to represent the amplitude of the profile deviations (normal vector to the nominal surface).

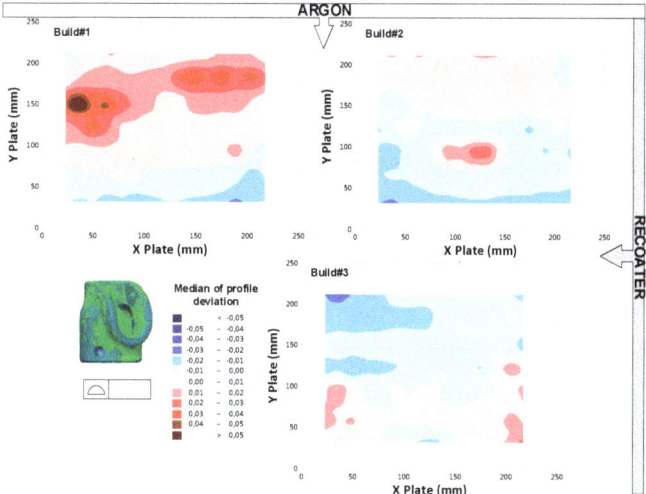

Figure 4. Contour plot of the profile deviation distribution using a median deviation of each part for all three builds.

The results of Analysis 1b are presented in Figure 5. Colors are brought about to distinguish the material withdrawal, when the feature is smaller than the nominal size in the least material condition (LMC) direction from the addition which is an increase from the nominal size in the maximum material condition (MMC) direction as in ASME Y14. 5.1 [33]. Black bubbles are placed where this difference was less than 1 µm.

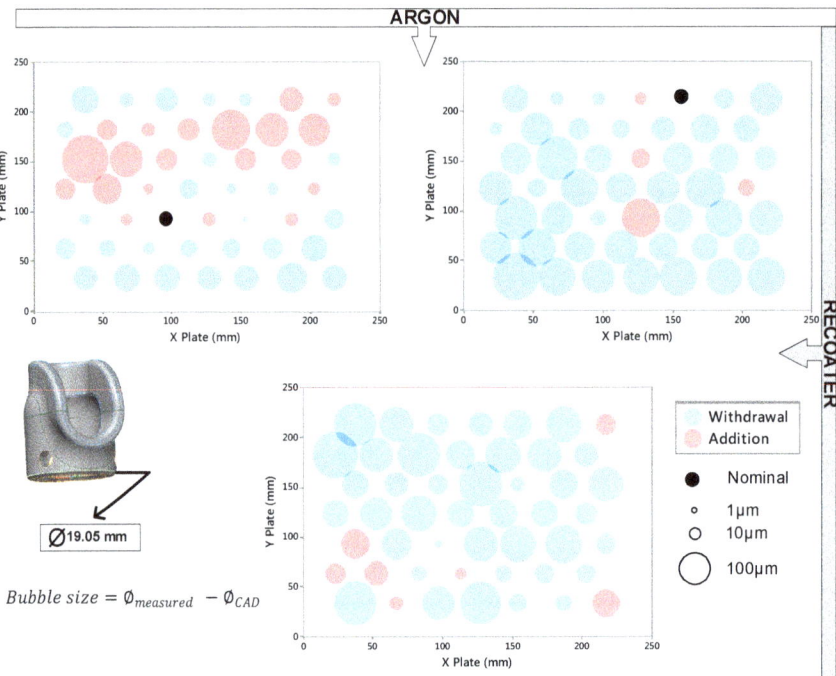

Figure 5. Bubble plot of the diameter deviation of each part of the three builds; the size of the bubble illustrates the absolute difference between the measured diameter and the nominal size of the part.

The results of Analysis 1c correlating the measured diameter (predictor) and the median profile deviations (response) are plotted in Figure 6. In Figure 6, the value of S is measured in units of the response variable and represents the standard distance data values from the regression line. For a given study, the better the equation predicts the response, the lower the S value. R-Sq represents the proportion of variation in the observed response values that is explained by the predictor variable, which is the measured diameter. Adjusted R-Sq(adj) is a modified R that has been adjusted for the number of terms in the model.

A basic statistical study was also conducted to evaluate the intra-build variation (Analysis 1d). The first objective of this analysis was the external diameter at a height of $z = 1.2$ mm extraction and characterization. The results are presented in Table 1. The second objective is the overall 3D profile deviations of each part, represented by the gap between the non-parametric quantiles $L_{1\%}$ and $L_{99\%}$ (Table 2).

Table 1. Descriptive statistics of the measured diameter for 49 parts (dimensions in mm).

Build	$\mu_\varnothing$	$StDev_\varnothing$	$Min_\varnothing$	$Median_\varnothing$	$Max_\varnothing$
#1	19.053	0.054	18.970	19.041	19.243
#2	19.017	0.025	18.964	19.015	19.108
#3	19.012	0.038	18.936	19.011	19.095

With μ = mean; $StDev$ = Standard deviation.

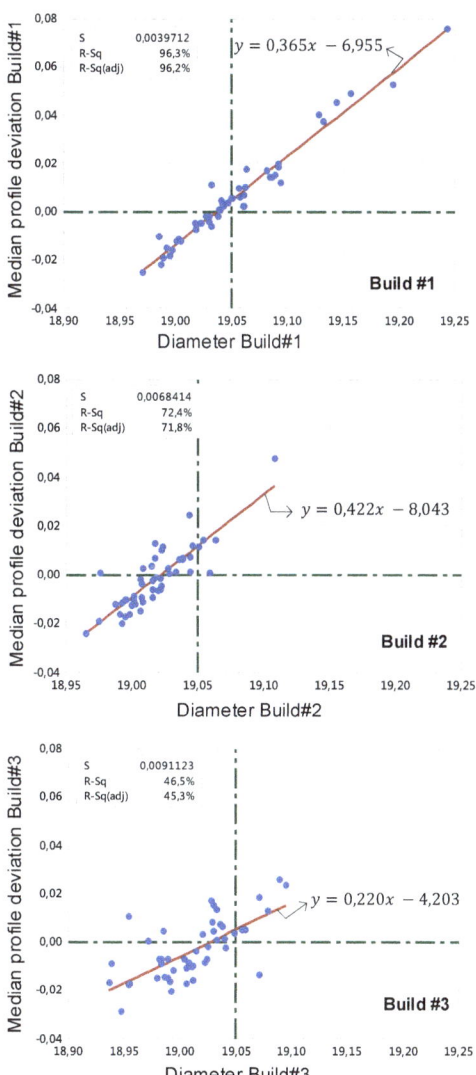

Figure 6. Correlation between the diameter deviation (predictor) and the profile deviation (response).

Table 2. Descriptive statistics of the measured Profile ⌂ ($L_{99\%} - L_{1\%}$) for 49 parts (dimensions in mm).

Build	μ ⌂	StDev ⌂	Min ⌂	Median ⌂	Max ⌂
#1	0.148	0.058	0.108	0.131	0.501
#2	0.152	0.023	0.124	0.149	0.276
#3	0.147	0.014	0.116	0.148	0.181

3.2. Inter-Build Variations

This study involves comparing the builds and quantifying and analyzing the differences. First of all, a visual comparison is carried out. For example, Figure 7 presents the overall 3D-profile deviations for Build #2 and Build #3, where the same color scale and parameters are used. This comparison reveals more material withdrawal in Build #3 than in Build #2 (more detailed discussion will be made in Section 4). Next, Figure 8 illustrates the results of Analysis 2a (KS test). Since the *p*-value is higher than 0.05 (α), no significant statistical differences between the CDFs of Build #2 and Build #3 can be reported (95% confidence level). The range of the inter-repeatability (Analysis 2b) for the 49 locations is 455 μm. The minimum part variation is 14 μm, and the maximum is 469 μm at a 95% confidence level, as will be shown in more detail in the next section.

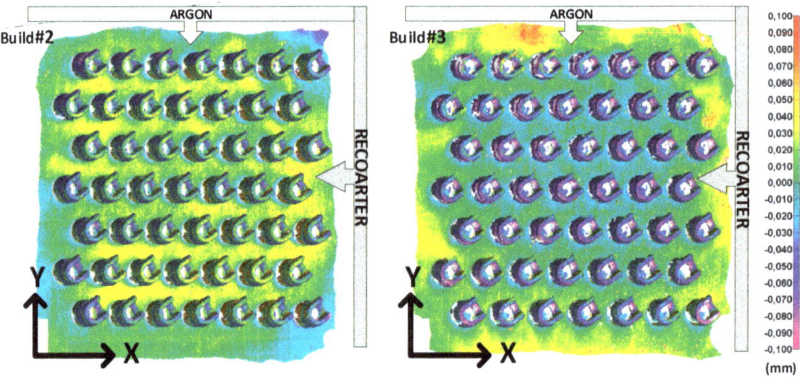

Figure 7. Overall 3D profile color deviation map for Build #2 and Build #3.

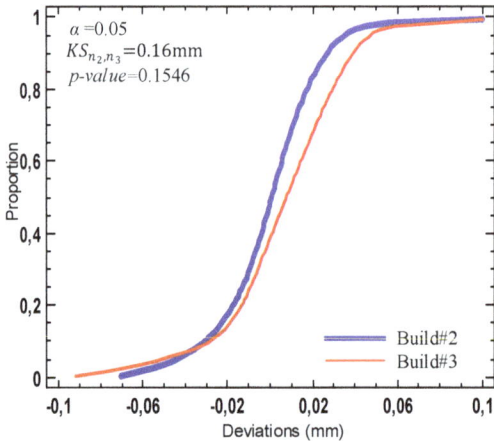

Figure 8. KS-test results for Build #2 and Build #3.

3.3. Capability

The capability study (Analysis 3) was performed on all 174 parts, and the results of this study are presented in Figure 9. Figure 9a illustrates the external diameter extraction and quantification, Figure 9b presents its non-parametric distribution, Figure 9c the distribution of the profile deviation of one part, with the capability interval highlighted, and Figure 9d shows the distribution of the capability intervals of 49 parts.

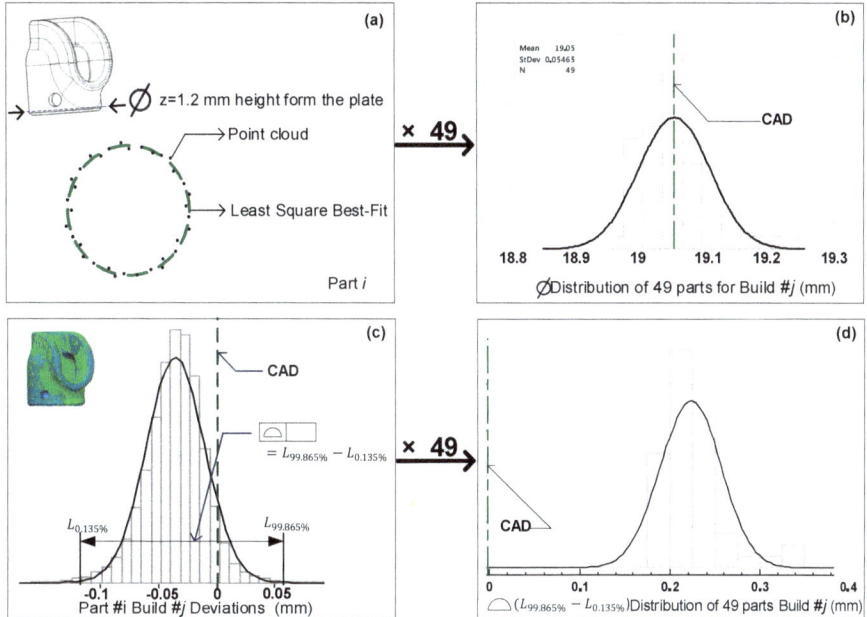

Figure 9. Capability and diameter deviation analyses: (**a**) Diameter quantification, (**b**) 49 parts' (one build) diameter distribution, (**c**) 3D profile deviation capability, and (**d**) 49 parts' (one build) 3D profile deviation capabilities distribution.

The results of Analysis 3a are presented in Figure 10, giving an overview of the capabilities (as in Equation (7)) over three builds for nine locations selected to uniformly cover the build space. Thus, for each of the selected part location, the capability (99.73%) and the 95% ($L_{97.5\%} - L_{2.5\%}$) intervals of profile deviations are provided for Build #1, Build #2, and Build #3. Table 3 presents the results of Analysis 3b for Builds #1, 2, 3 and for the overall 147 parts. It also reveals that the 3D profile deviation capability interval for the 147 parts falls within 228 µm at the 99.73% level.

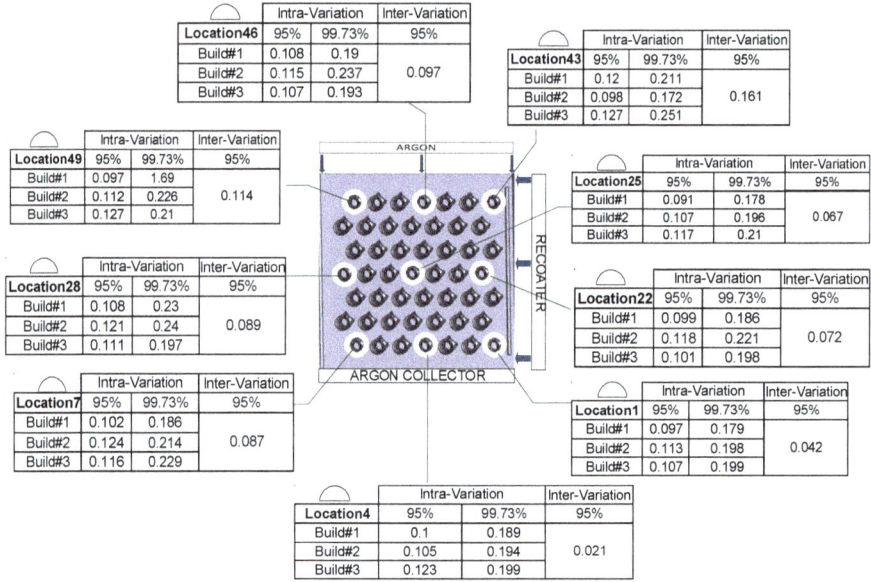

Figure 10. Intra and inter-variation of part profile deviation study (mm).

Table 3. 3D profile deviation (mm) and equivalent IT grade (International Tolerance Grade defined in ISO 286).

Build	$\mu_\triangle$	$StDev_\triangle$	95%	97.73%
#1	0.005	0.034	0.136	0.240
#2	0.000	0.032	0.127	0.225
#3	−0.002	0.030	0.121	0.191
Overall	0.001	0.032	0.128 (IT 11)	0.228 (IT 12)

4. Discussion

After the first build, neither the second nor the third build showed any similarity in terms of the distribution (pattern) of the 3D profile deviations in the manufacturing chamber. Globally, the deviation values are in the same range, but their distribution in the chamber is not repeating. We can then conclude there is no specific pattern of geometric deviations on the chamber for LPBF process with an EOS M280. The measured range of the intra-build means variations are 0.100 mm for the first build, 0.071 mm for the second, and 0.054 mm for the third build. The inter-build variation range is 0.104 mm. The intra-build variations are practically constant even if their distribution on the build plate is not similar. The observation of Figure 7 highlights more withdrawal in Build #3 than Build #2 (Figure 11a). However, since the magnitude of the differences between the two builds is lower than the measurement equipment uncertainty which is ±5 µm, we cannot really conclude on the absence of any significant difference between these builds. The range of the intra-build diameter (Ø19.05 mm) variations at z = 1.2 mm is 0.273 mm for the first build, 0.144 mm for the second, and 0.159 mm for the third build. The overall diameter deviation variation range is 0.307 mm (Figure 11b) which corresponds to an equivalent IT Grade IT 13. The 3D profile deviation behavior of the 147 parts falls within 128 µm at a 95% level, which corresponds to an IT 11. The 3D profile capability interval (99.73%) for the process is 228 µm, which is an IT 12 equivalent, comparable to turning and milling process tolerance.

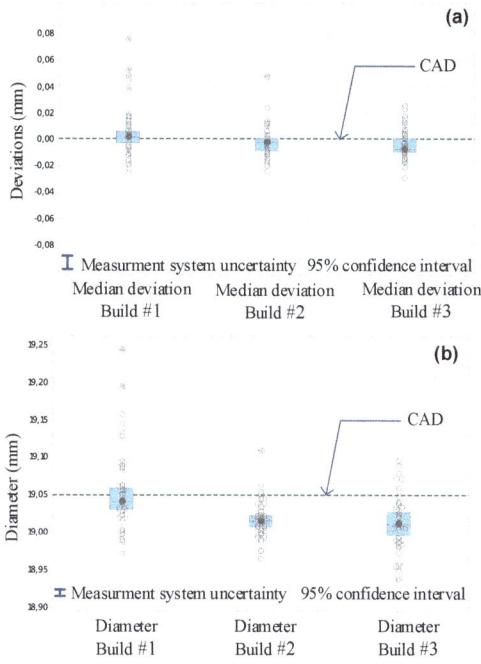

Figure 11. Box plot of the profile deviation (**a**) and diameter deviation (**b**).

5. Conclusions

This paper presents a metrological investigation carried out on 147 typical aerospace tooling components built in three print jobs using an AlSi10Mg powder and an EOS M280 LPBF system. The investigations were limited to the overall 3D profile and diameter deviation studies, specifically to their repartition in the build chamber. No significant statistical differences were revealed between the 49 locations over the three builds, and the deviation distribution in the build chamber appeared to be non-repeatable. However, inspection of part external diameters reveals a correlation between this feature and the overall 3D profile deviation. In fact, it was shown that the magnitude of these deviations is in the same range as the measurement equipment uncertainty, which is ±5 µm. Further studies with different geometries, such as cylinders, holes, cubes, and cones, could be promising.

The results of this study, and of the upcoming ones, will have a positive impact on increasing the competitiveness of the LPBF process. The findings of the study can also be directly applied to high technology industries, such as aerospace and automotive sectors, planning to use the metallic AM technology in their production cycle.

Author Contributions: The project objectives and methodology were proposed by A.T. and V.B. The specimen fabrication, scanning and data treatment were carried out by F.Z. with the help of A.T., V.B. and A.A. The article was written by F.Z. and revised by A.T., V.B. and A.A.

Funding: This research received no external funding.

Acknowledgments: The authors would like to thank the Natural Sciences and Engineering Research Council of Canada (NSERC) and École de technologie supérieure (ETS) for their supports. The authors are thankful to Joel Grignon, Anatoli Timercam, Morgan Letteneur, and Jean-René Poulin who assisted in this research.

Conflicts of Interest: The authors declare no conflict of interest.

References

1. ASTM-ISO. *Standard Guide for Additive Manufacturing—General Principles—Requirements for Purchased AM Parts*; ASTM52901-16; ASTM International: West Conshohocken, PA, USA, 2016.
2. ASTM-ISO. *ISO/ASTM 52900: 2015 Additive Manufacturing—General Principles—Terminology*; ASTM F2792-10e1; ASTM International: West Conshohocken, PA, USA, 2012.
3. Wohlers, T.; Caffrey, T. *Wohlers Report 2014: Additive Manufacturing and 3D Printing State of the Industry*; Annual Worldwide Progress Report; Wohlers Associates Inc.: Fort Collins, CO, USA, 2017.
4. Shiomi, M.; Osakada, K.; Nakamura, K.; Yamashita, T.; Abe, F. Residual stress within metallic model made by selective laser melting process. *CIRP Ann. Manuf. Technol.* **2004**, *53*, 195–198. [CrossRef]
5. Wang, X. Calibration of shrinkage and beam offset in SLS process. *Rapid Prototyp. J.* **1999**, *5*, 129–133. [CrossRef]
6. Zhu, H.; Lu, L.; Fuh, J. Study on shrinkage behaviour of direct laser sintering metallic powder. *Proc. Inst. Mech. Eng. Part B* **2006**, *220*, 183–190. [CrossRef]
7. Raghunath, N.; Pandey, P.M. Improving accuracy through shrinkage modelling by using Taguchi method in selective laser sintering. *Int. J. Mach. Tools Manuf.* **2007**, *47*, 985–995. [CrossRef]
8. Islam, M.N.; Sacks, S. An experimental investigation into the dimensional error of powder-binder three-dimensional printing. *Int. J. Adv. Manuf. Technol.* **2016**, *82*, 1371–1380. [CrossRef]
9. Senthilkumaran, K.; Pandey, P.M.; Rao, P. New model for shrinkage compensation in selective laser sintering. *Virtual Phys. Prototyp.* **2009**, *4*, 49–62. [CrossRef]
10. Galovskyi, B.H.T. Testing Workpieces for Selective Laser Sintering. In Proceedings of the ASPE 2015 Spring Topical Meeting, Golden, CO, USA, 8–10 July 2015; pp. 89–94.
11. Van Bael, S.; Kerckhofs, G.; Moesen, M.; Pyka, G.; Schrooten, J.; Kruth, J.-P. Micro-CT-based improvement of geometrical and mechanical controllability of selective laser melted Ti6Al4V porous structures. *Mater. Sci. Eng. A* **2011**, *528*, 7423–7431. [CrossRef]
12. Silva, D.N.; De Oliveira, M.G.; Meurer, E.; Meurer, M.I.; da Silva, J.V.L.; Santa-Bárbara, A. Dimensional error in selective laser sintering and 3D-printing of models for craniomaxillary anatomy reconstruction. *J. Cranio-Maxillo-Fac. Surg.* **2008**, *36*, 443–449. [CrossRef] [PubMed]
13. Vanderesse, N.; Ky, I.; González, F.Q.; Nuño, N.; Bocher, P. Image analysis characterization of periodic porous materials produced by additive manufacturing. *Mater. Des.* **2016**, *92*, 767–778. [CrossRef]
14. Kruth, J.-P. Material incress manufacturing by rapid prototyping techniques. *CIRP Ann. Manuf. Technol.* **1991**, *40*, 603–614. [CrossRef]
15. Lart, G. Comparison of rapid prototyping systems. In Proceedings of the First European Conference on Rapid Prototyping, University of Nottingham, Nottingham, UK, 6–7 July 1992; pp. 6–7.
16. Grimm, T. Fused deposition modelling: A technology evaluation. *Time Compress. Technol.* **2003**, *11*, 1–6.
17. Castillo, L. *Study about the Rapid Manufacturing of Complex Parts of Stainless Steel and Titanium*; TNO Report with the Collaboration of AIMME; TNO: Delft, The Netherlands, 2005.
18. Abdel Ghany, K.; Moustafa, S. Comparison between the products of four RPM systems for metals. *Rapid Prototyp. J.* **2006**, *12*, 86–94. [CrossRef]
19. Dimitrov, D.; Van Wijck, W.; Schreve, K.; De Beer, N. Investigating the achievable accuracy of three dimensional printing. *Rapid Prototyp. J.* **2006**, *12*, 42–52. [CrossRef]
20. Hanumaiah, N.; Ravi, B. Rapid tooling form accuracy estimation using region elimination adaptive search based sampling technique. *Rapid Prototyp. J.* **2007**, *13*, 182–190. [CrossRef]
21. Cooke, A.L.; Soons, J.A. Variability in the geometric accuracy of additively manufactured test parts. In Proceedings of the 21st Annual International Solid Freeform Fabrication Symposium, Austin, TX, USA, 9–11 August 2010; pp. 1–12.
22. Moylan, S.; Slotwinski, J.; Cooke, A.; Jurrens, K.; Donmez, M.A. Proposal for a standardized test artifact for additive manufacturing machines and processes. In Proceedings of the 2012 Annual International Solid Freeform Fabrication Symposium, Austin, TX, USA, 6–8 August 2012; pp. 6–8.
23. Minetola, P.; Iuliano, L.; Marchiandi, G. Benchmarking of FDM machines through part quality using IT grades. *Procedia CIRP* **2016**, *41*, 1027–1032. [CrossRef]
24. Fahad, M.; Hopkinson, N. A new benchmarking part for evaluating the accuracy and repeatability of Additive Manufacturing (AM) processes. In Proceedings of the 2nd International Conference on Mechanical, Production and Automobile Engineering (ICMPAE 2012), Singapore, 28–29 April 2012; pp. 28–29.

25. Teeter, M.G.; Kopacz, A.J.; Nikolov, H.N.; Holdsworth, D.W. Metrology test object for dimensional verification in additive manufacturing of metals for biomedical applications. *Proc. Inst. Mech. Eng. Part H* **2015**, *229*, 20–27. [CrossRef] [PubMed]
26. Ferrar, B.; Mullen, L.; Jones, E.; Stamp, R.; Sutcliffe, C. Gas flow effects on selective laser melting (SLM) manufacturing performance. *J. Mater. Process. Technol.* **2012**, *212*, 355–364. [CrossRef]
27. Aidibe, A.; Tahan, A.; Brailovski, V. Metrological investigation of a selective laser melting additive manufacturing system: A case study. *IFAC-PapersOnLine* **2016**, *49*, 25–29. [CrossRef]
28. Rebaioli, L.; Fassi, I. A review on benchmark artifacts for evaluating the geometrical performance of additive manufacturing processes. *Int. J. Adv. Manuf. Technol.* **2017**, *93*, 2571–2598. [CrossRef]
29. Sing, S.L.; Wiria, F.E.; Yeong, W.Y. Selective laser melting of lattice structures: A statistical approach to manufacturability and mechanical behavior. *Robot. Comput. Integr. Manuf.* **2018**, *49*, 170–180. [CrossRef]
30. Calignano, F. Investigation of the accuracy and roughness in the laser powder bed fusion process. *Virtual Phys. Prototyp.* **2018**, *13*, 97–104. [CrossRef]
31. Wilcox, R. Kolmogorov–Smirnov Test. In *Encyclopedia of Biostatistics*; John Wiley & Sons, Ltd.: New York, NY, USA, 2005.
32. Mooney, C.Z. *Monte Carlo Simulation*; Sage Publications: Thousand Oaks, CA, USA, 1997; Volume 116.
33. American Society of Mechanical Engineers. *Mathematical Definition of Dimensioning and Tolerancing Principles: ASME Y14. 5.1 M-1994*; American Society of Mechanical Engineers: New York, NY, USA, 1995.

© 2018 by the authors. Licensee MDPI, Basel, Switzerland. This article is an open access article distributed under the terms and conditions of the Creative Commons Attribution (CC BY) license (http://creativecommons.org/licenses/by/4.0/).

Article

Effect of Build Orientation on the Microstructure and Mechanical Properties of Selective Laser-Melted Ti-6Al-4V Alloy

Patrick Hartunian and Mohsen Eshraghi *

Department of Mechanical Engineering, California State University, Los Angeles, CA 90032, USA; phartun@calstatela.edu
* Correspondence: mohsen.eshraghi@calstatela.edu; Tel.: +1-323-343-5218

Received: 9 September 2018; Accepted: 10 October 2018; Published: 12 October 2018

Abstract: One of the challenges of additive manufacturing (AM) technology is the inability to generate repeatable microstructure and mechanical properties in different orientations. In this work, the effect of build orientation on the microstructure and mechanical properties of Ti–6Al–4V specimens manufactured by selective laser melting (SLM) was studied. The samples built in the Z orientation showed weaker tensile strength compared to the samples built in X, and Y orientations. Samples built in X and Y orientations exhibited brittle fracture features in areas close to the substrate and ductile fracture features in the area farther from the substrate. Defects including pores, cracks, and unmelted/partially-melted powder particles contributed to lower tensile and fracture toughness properties in different orientations.

Keywords: selective laser melting; build orientation; Ti–6Al–4V; microstructure; mechanical properties; surface roughness

1. Introduction

Additive manufacturing (AM) is defined as a process of "joining materials to make objects from 3D model data, usually layer upon layer, as opposed to subtractive manufacturing methodologies" [1]. AM is suitable for small- and medium-part production and enables design flexibility and freedom in comparison with conventional manufacturing processes where complex geometries and net shape products are desired. However, some challenges still exist, such as limitations in part size, production number, and repeatability of material properties [2]. One of the challenges facing AM technology is the ability to achieve identical microstructure and mechanical properties in different build orientations. Numerous studies have been conducted to improve the quality of the additively manufactured parts and ensure repeatable material properties.

The effect of altered process parameters on AM-fabricated Ti–6Al–4V alloys was investigated by Gong et al. [1]. It was found that energy density has a significant effect on defects and porosity of the samples generated by powder bed fusion (PBF) [1]. The microstructure and mechanical properties of Ti–6Al–4V tensile specimen fabricated by selective laser melting (SLM) and electron beam melting (EBM) were studied by Rafi et al. [3]. The α′ martensitic phase was the microstructure of SLM specimen due to processing parameters and cooling rate. The microstructure of EBM-processed samples consisted of primary α and a small amount of β phase due to elevated temperature in the build chamber [3]. The main types of microstructure for titanium alloys were reported as lamellar α within large β grain, which forms during slow cooling rates and can be characterized by low ductility, moderate fatigue properties along with good creep and crack resistance. The second type is equiaxed two phase α + β for fast cooling processes with better balance of strength and ductility along with fatigue properties [4]. The formation of α and β microstructure in as-fabricated SLM Ti–6Al–4V was simulated by variation of process parameters, but the typical SLM microstructure was martensitic α′ based on scanning speed and rapid cooling [5]. Heat treatment can decompose α′ into α and β phases. SLM intrinsic heat treatment converts α′ martensitic phase into α + β microstructure during sintering. The intensified treatment employs tight hatch spacing along with high energy density and elevated temperature of the platform. However, high energy density and elevated temperature may result in void defects [6]. Rapid heat transfer, melt pool flow and geometry influence the grain size and microstructure of the printed material [7–10]. The way in which an increase in the current and frequency of the laser elevates the density and thickness of the specimen was studied by Fatemi et al. Higher scanning speed has the opposite effect on layer thickness. Maximum density was obtained with high current, frequency and decreasing scanning speed [11]. Intrinsic heat treatment during SLM facilitates transformation of the α′ martensitic phase into α and β microstructures. The intensified treatment employs tight hatch distance along with high energy density and elevated temperature to reduce the cooling rate [6]. The Marangoni convection in the melt pool may also make the pool unsteady, despite constant scanning velocity [7,12]. Melt pool geometry is influenced by scanning speed, and laser power among other factors. The rapid solidification behavior is due to large thermal gradients and high thermal conductivity of metallic alloys. As a result, fine grains with refined microstructure are generated. Also, the solubility of a solid may be extended and chemical homogeneity may increase along with crystalline, quasicrystalline, and amorphous metastable phases [13,14].

One of the common defects resulted from SLM processes is porosity. Formation of defects such as porosity is directly related to the laser beam and powder interactions during the process. Interaction of the energy source in SLM and EBM processes with powder particles results in extreme temperature and massive liquid formation. During solidification gaseous bubbles float over the liquid by Marangoni flow. Trapped bubbles in the solidified region form pores. Porosity can be reduced by optimizing the hatch spacing, laser power and scan speed [15,16]. Porosity can be powder-induced or process-induced. Powder-induced defects include spherical gaseous voids inside the powder. The powder is held in place by gravitational force and is under the influence of a powder distribution mechanism, rapid temperature change, and capillary forces. There is no pressure source to compact the powder particles or hold them closer. Porosity may also consist of large irregular unmelted powder zones, shrinkage pores due to lack of enough powder in the interdendritic zones, and spherical pores due to gas entrapped in the part [7]. Process-induced porosity is due to insufficient energy to melt the powder completely. Lack of fusion can be recognized where unmelted powder particles are observable near the pores and shrinkage porosity is due to lack of powder in the melt pool. Spatter ejection is another phenomenon induced by high beam power, where the melt pool boils and drives the molten material out of melt pool via a convection process [17]. Balling is another defect that results from inability of the molten material to connect effectively with previous layers. Therefore, the surface of the material becomes rough and porous with bead-shaped tracks. Material properties and processing parameters are to be blamed for the problem [7]. Low scanning speed and high energy density lead to higher melting pool temperature and viscosity of the liquid. The splashing of liquid droplets on the solidified surface is another reason for balling [18]. The excessive thermal gradient between melt pool and its surroundings is also a source of cracks on the surface and core [7]. Cracking may occur during solidification and depends on the material's dendritic, cellular or planar solidification nature [17]. The orientation-dependent microstructure, defects, and texture influence the tensile properties, but it is more critical in fracture properties. Post-processing can mitigate the process-dependent defects [16]. High laser power and low scan speed stabilizes the melt pool geometry significantly [18]. The effect of build orientation on the mechanical properties of Ti–6Al–4V alloy processed by SLM was studied by Simonelli et al. [19]. The directionality of prior β grain boundaries to the external axial loading affected the fracture mechanism and crack growth in the parts [19]. Another study found that the fracture toughness was comparably higher when build layers were perpendicular to crack growth direction [20]. Recently, Barriobero-Vila et al. [21] developed a new Ti alloy by exploiting metastability around peritectic and peritectoid reactions. Their findings promise a decrease in anisotropy of as-built and heat-treated Ti alloys for AM [21].

The purpose of this research is to investigate the effect of build orientation on the mechanical properties and microstructure of the titanium Ti–6Al–4V alloy manufactured by selective laser melting. The research involved tensile, and fracture toughness tests of samples printed in three different orientations. Metallography, hardness, and surface roughness analysis were also performed. The microstructure and fracture surfaces of the samples were studied using optical and scanning electron microscopy techniques.

2. Materials and Methods

To characterize orientation-based microstructure and mechanic properties of SLM Ti–6Al–4V alloy, rectangular samples were printed by Renishaw plc. AM250 200 W platform with 70 µm spot size. The system uses high stability ytterbium fiber lasers, guided through an optical module to deliver a positioning accuracy of ±25 µm across the working area. The power is delivered via a point-by-point exposure methodology.

The process parameters used for printing samples are listed in Table 1. General print settings include layer thickness, point distance, exposure time, power, focus and hatch distance.

Table 1. Process parameters used for manufacturing test samples.

Environment	Argon/Nitrogen
Exposure time	50 µs
Focal point	75 µm
Laser power	200 W
Layer thickness	30 µm
Operational temperature	170 °C
Hatch spacing	75 µm

Samples were fabricated in three perpendicular orientations without any support. To prevent porosity, the melt pool was overlapped by sufficient distance. The powder used for printing samples was Ti–6Al–4V ELI-0406 alloy produced by Renishaw plc. with particle sizes ranging between 15 µm and 45 µm [22]. The powder bed orientation and setup for printed samples are shown in Figure 1.

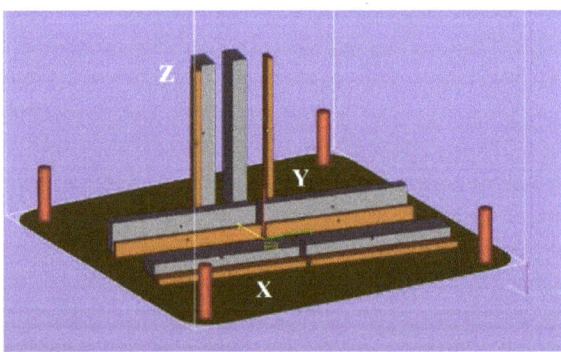

Figure 1. Sample geometry; print orientation and setup in isometric view.

The tensile and fracture toughness build orientations are designated according to nomenclature in Figure 1 by X, Y, and Z. The first letter in the sample ID, T or F, represents tensile and fracture toughness samples, respectively. The orientation nomenclature of the tensile and fracture toughness samples are represented in Table 2.

Table 2. Tensile and fracture toughness sample nomenclature.

Orientation	Tensile	Fracture
XYZ (X)	TX	FX
XZY (Y)	TY	FY
ZXY (Z)	TZ	FZ

All samples were heat treated after the SLM process. The heat treatment of the samples was performed under an argon gas environment with an initial ramp to 850 °C over 110 min, held at that constant temperature for 60 min, then furnace-cooled to 350 °C before turning off argon flow.

The tensile samples were generated in two sets of oversized rectangular shape in three Cartesian build directions. The samples were considered 0.0625" (1.58 mm) larger to comply with the dimensions shown in Figure 2 after machining process. Tensile samples were machined per ASTM E8 standard. The rough dimensions of samples were about 4.062 × 0.400 × 0.187 inches (102 × 10.2 × 4.8 mm).

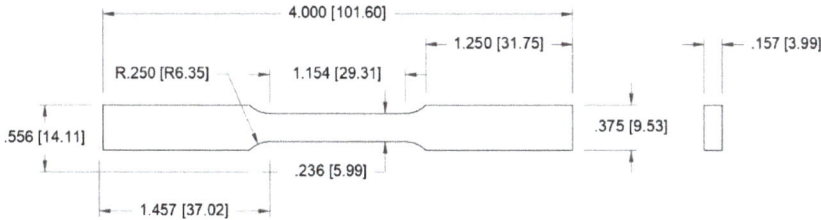

Figure 2. Tensile specimen dimensions in inches (mm).

The build orientations comply with ASTM nomenclature. The tensile test was performed using a universal tensile test machine. The tensile test was conducted per ASTM E8 standard [23]. The fracture toughness samples were fabricated in two sets of rectangular shape, in three Cartesian build orientations. The fracture toughness samples were tested in as-built condition. However, notches were machined in accordance with ASTM E399 standard [24], as displayed in Figure 3.

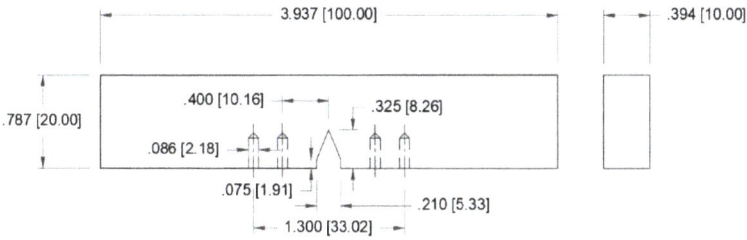

Figure 3. Fracture toughness sample dimensions in inches (mm).

In order to investigate as-built and fractured surfaces, scanning electron microscopy (SEM) was undertaken using Zeiss Ultra 55 FESEM. In addition, electron dispersive spectroscopy (EDS) was performed by Oxford EDAX/EBSD equipment on samples in order to verify the consistency of chemical composition in all build orientations. Instron Wilson 2000 hardness tester was used for hardness measurements, and optical microscopy was performed using Nikon Epiphot 300 microscope to analyze the microstructure. Furthermore, Wyco NT9100 optical surface profilometery was employed to analyze the surface roughness of the as-built samples.

3. Results

EDS analysis was performed on the finished surface of one set of the tensile samples, in two random locations. Measured compositions by EDS were in compliance with Renishaw Ti–6Al–4V ELI-0406 stated composition. The stated composition was up to 90% titanium mass fraction alloyed with up to 6.75% aluminum and up to 4.5% vanadium and other minor elements [22]. The results reflect no major differences in chemical composition at different orientations after the SLM process. No dilution, diffusion or evaporation was observable in the results.

3.1. Tensile Test

Table 3 shows the average tensile properties, which include yield strength, ultimate tensile strength, strain percentage, and modulus of elasticity.

Table 3. Tensile test results in different build orientations.

Build Orientation	Yield Strength MPa (ksi)	Ultimate Strength MPa (ksi)	Strain %	E MPa (psi)
TX	1002.15 (145.35)	1023.9 (148.5)	1.13	97,216 (14.1 × 10^6)
TY	981.81 (142.40)	1018.3 (147.7)	1.30	97,216 (14.1 × 10^6)
TZ	868.05 (125.90)	872.9 (126.6)	0.98	95,148 (13.8 × 10^6)

The strain % in Z orientation is considerably lower than X and Y orientations. TZ samples failed shortly after reaching ultimate strength, showing minimal plastic deformation. The comparable low strain % and relatively close yield and ultimate tensile strength values can be associated with the brittle fracture in the Z orientation, as shown in the SEM micrographs presented below.

The fractured surfaces of tensile samples were investigated by SEM. The TZ orientation, Figure 4a, shows a brittle planar fracture perpendicular to the building orientation, with smoother texture compared to other build orientations. The planar fracture indicates possible interlayer failure due to lack of strong bonding between successively deposited layers. This may have happened due to insufficient laser exposure or high scan speed that was unable to melt the deeper layers. The presence of partially melted and unmelted powder particles also suggests the interlayer fracture. A ductile fracture with representative dimples is also observable in certain areas as shown in Figure 4b. The fracture surface includes cleavage, dimples from ductile failure, voids and unmelted powder particles as depicted in Figure 4b,c.

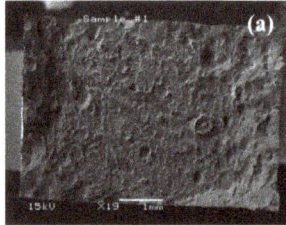 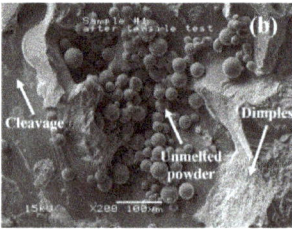 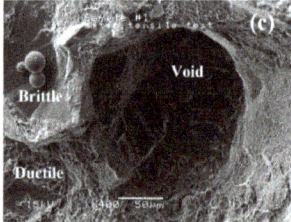

Figure 4. Scanning electron microscope (SEM) fractographs of TZ sample: (a) the entire fracture surface; (b) unmelted powder particles; (c) mixed brittle and ductile fracture features.

The TY sample surfaces demonstrated a combination of brittle and ductile fracture. As shown in Figure 5, the areas closer to substrate show a brittle fracture, while ductile fracture characteristics can be identified in the other areas. This can be justified by cooling rate behavior during the process. The initial layers go through a faster cooling due to the proximity to the cold substrate. The cooling rate decreases as subsequent layers are deposited on top of hot printed layers.

Figure 5. SEM fractographs of TY sample: (**a**) fracture features with respect to build orientation; (**b**) interlayer fracture.

The TX fracture surfaces also display a mixed brittle and ductile behavior. Similar to the TY sample, but in a perpendicular orientation, the starting layers above the substrate in the TX sample, Figure 6a, exhibited a brittle fracture due to a high cooling rate while the remaining section shows a ductile fracture. Partially melted and unmelted powders along with voids are also visible in all samples as shown in Figure 6b.

Figure 6. SEM fractographs of TX sample: (**a**) fracture features with respect to build orientation; (**b**) unmelted powder particles and their sizes.

Figure 6b illustrates the accumulation of unmelted particles in the TX sample. The powder size is consistent with the initial powder size. The unmelted powder collection near the smooth surface may be due to a Marangoni flow that rejected powder particles from the molten pool. The variation of temperature in different melt pool locations creates a flow of molten material from the center of the melt pool to the surrounding area. The molten material at the edge moves to the bottom of the pool while the dislocated fluid moves to the top due to buoyancy force and variation of density between hot and cold regions, causing a Marangoni convection [25].

3.2. Fracture Toughness

Fracture toughness measures a material's resistance to the extension of a crack. Orientation-dependent differences in the microstructure, texture, and defects contribute to differences in mechanical properties. Defects, pores, and unmelted powder particles contributed to inconsistent fracture toughness results. The average fracture toughness properties are reported in Table 4.

The FY orientation shows the highest fracture toughness (K_q) values. In the FY samples, additively manufactured layers and interlayer cracks are perpendicular to the notch orientation, preventing the notch crack from propagation. Since the interlayer cracks in the FZ samples are parallel to the notch, FZ orientation is expected to have the lowest fracture toughness properties. However, FX orientation shows the lowest values. Relatively higher hardness values in the Z orientation may have contributed to better fracture toughness properties. Fracture surfaces exhibited a mixed mode of brittle cleavage and ductile fracture, as shown in Figure 7. The schematics on the lower right corner of Figure 7 show the print orientation and arrangement of the layers. Areas of perfect solidification were characterized as transgranular ductile dimple fractures from the coalescence of microvoids. Fine dimples at the tensile fracture surface were indications of plastic deformation. Defects were observed on the surfaces of fracture toughness samples, which consisted of isolated porosities, voids, cracks, and unmelted regions. Lack of laser power, pulse frequency along with high scanning speed may contribute to large areas of voids and unmeted powder [26].

Table 4. Ti–6Al–4V fracture toughness results.

Orientation	P_Q Kgf (lbs)	Kq MPa$\sqrt{m}$ (psi$\sqrt{in}$)	Kq Std. Deviation
FX	1073.6 (2367)	55.6 (50,953)	0.99 (1438.25)
FY	1031.5 (2274)	57.8 (55,507)	1.06 (3107.03)
FZ	992.0 (2187)	61.6 (52,841)	2.54 (2285.37)

Areas of unmelted powder on top of smooth solidified material might be generated by Marangoni flow during the melting process. Marangoni flow can move the molten material away from the center of the melt pool (Figure 8). Optimization of the processing parameters can potentially eliminate some of these defects [26]. Similar pores were observed in samples printed by EONSINT M270 [27] as compared in Figure 8. Both SEM micrographs show ductile fracture with pores and unmelted powders.

Figure 7. SEM micrograph of the fracture surfaces: (**a**) FX; (**b**) FY; (**c**) FZ samples. The schematics show the print orientation and arrangement of the layers.

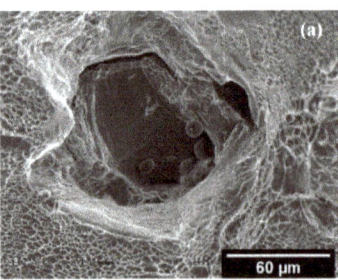 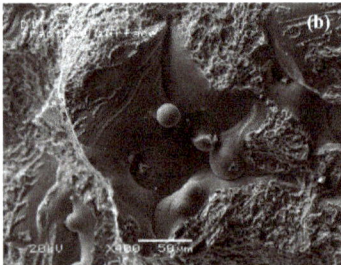

Figure 8. SEM micrograph of the fracture surface: (**a**) EONSINT M270 [27]; (**b**) AM250 (current study).

Table 5 provides Kq toughness values of PBF Ti–6Al–4V samples for as-built, stress-relieved, hot isostatic pressing (HIP), and heat-treated conditions as reviewed by Lewandowski and Seifi [16]. The orientations listed in the table comply with ASTM standard number 52921 [28]. ASTM standard 59921 explains the terminology for AM coordinate systems and test methodologies. Z designates the build direction. X is parallel to the machine front and perpendicular to Z. Y is perpendicular to the Z and X axes, with a right-hand positive coordinate definition. The first letter is the axis parallel to the peak dimension. The second letter determines the second larger dimension [29].

Table 5. PBF Ti–6Al–4V fracture toughness data comparison.

Machine	Condition	Orientation	Kq (MPa√m)	Kq (psi√in)	Reference
SLM	As built	XY	28 ± 2	25,481	[30]
		XZ	23 ± 1	20,931	
		ZX	16 ± 1	14,561	
	Stress relieved	XY	28 ± 2	25,481	
		XZ	30 ± 1	27,301	
		ZX	31 ± 2	28,211	
	Heat treated	XY	41 ± 2	37,312	
		XZ	49 ± 2	44,592	
		ZX	49 ± 1	44,592	
SLM MTT250	As built	XY	66.9 ± 2.6	60,882	[20]
		XZ	64.8 ± 16.9	58,971	
		YZ	41.8 ± 1.7	38,040	
SLM	As built	ZX	52.4 ± 3.48	47,686	[31]
EOS M280	As built	XY	37.5 ± 5	34,126	[32]
	HIP	XY	57.8 ± 5	52,600	
	Heat-treated	XY	86.3	78,537	
SLM	Heat-treated	XY	55.6 ± 1.0	50,953	Current Study
AM250		XZ	57.8 ± 1.1	55,507	Current Study
		ZX	61.6 ± 2.5	52,841	Current Study

It should be noted that the properties of the EOS M280 samples were significantly different in as-built and HIP PBF cases [16]. The EOS M280 samples heat treatment and HIP were rather complicated. The heat treatment performed was the recrystallization anneal at 950 °C for an hour twice, furnace-cooled, and air-cooled for the first and second processes, respectively. It was then heated to 700 °C for an hour and air-cooled followed by heating to 1030 °C for one hour, air-cooled and heated to 630 °C and air-cooled. For HIP, the samples were heated to 915 °C at 1000 bar isostatic pressure, two-hour holding period, and furnace-cooled at 11 °C/min [32]. The fracture toughness results for AM250 samples were comparable with EOS M280 HIP samples, but the fracture toughness properties were lower than heat-treated EOS M280 samples.

3.3. Metallography

The tensile samples were prepared for metallography to observe the microstructures in all three build orientations at different locations. The polished sample surfaces were swabbed with Keller etchant for about 6 s, and then neutralized and studied by optical microscopy. At the temperatures lower than β transus the alloy is a mixture of α and β phases. At high cooling rates the β transforms into martensitic α' phase. The α' phase might completely dominate the microstructure based on the cooling rate in the SLM process. However, the heat-treatment process transforms the martensitic α' phase into α and β phases. The metallography in all surfaces revealed α + β microstructure with no major differences in different orientations and magnifications. The microstructure in all build orientation was almost similar due to the heat treatment performed on the samples. Heat treatment above 600 °C coarsens partial martensitic α' plates into the laminar α + β structure. At β transus temperature, around 1000 °C, the coarsening is comparably higher. The coarse martensitic α' improves the mechanical properties, especially the ductility. The optimal heat treatment ranges from 850 °C to 950 °C meet standard specifications. In this study, heat treatment was performed under an argon gas environment with an initial ramp to 850 °C over 110 min, then held at 850 °C for 60 min, and finally furnace cooled to 350 °C before turning off the argon flow. Comparison of the microstructure in Figure 9 revealed that AM250 sample microstructure was similar to Vrancken et al. [33] samples heat-treated at 850 °C for two hours followed by air-cooling.

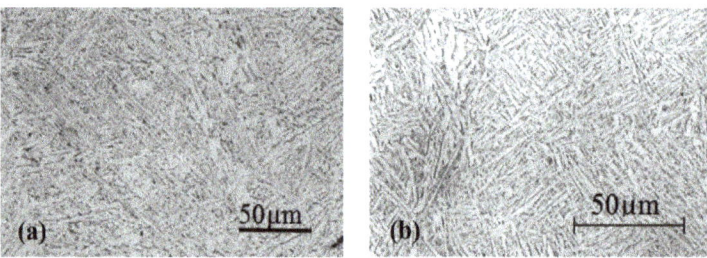

Figure 9. Comparison of microstructures in: (**a**) current study; (**b**) Vranken et al. [33].

3.4. Hardness Test

Hardness test was performed on all of the tensile samples. Hardness values varied between 29 to 38.6 Rockwell C scale (HRC), which were equal or superior to the fully annealed material, and comparable or lower than wrought coarse α and equiaxed α/β material [26]. Poonda et al. [34] performed hardness test on fully annealed Ti–6Al–4V AM samples. The hardness values for AM samples ranged from 25 to 30 HRC with an average of 26.675 HRC [34]. Table 6 compares the average hardness on bottom and top layers of the tensile samples. A declining trend was observed in hardness measurements from the bottom to the top. While the average hardness at the bottom of TZ samples was 37.67 HRC, the average hardness at the top layers was measured to be 30.10 HRC. A similar trend is observed for TX and TY samples. Higher hardness values were expected on starting layers compared to the remaining layers due to high cooling rates at the areas closer to the substrate. The hardness decreases as the distance from the substrate increases. However, the TZ sample shows a comparably higher hardness even at the top. The reason may be that when printing the top layers of TZ sample, the samples in X and Y orientations are already printed and the small cross section of TZ sample is surrounded by plenty of cold unmelted powder, which provides a high cooling rate. The defects present on the surface may have contributed to the variable hardness values as well [16]. The lower hardness results may be due to the presence of voids, unmelted powder particles and cracks. Optical microscopy of the samples revealed surface defects. For example, Figure 10 shows the interlayer defects perpendicular to build orientation in a TZ sample. The arrow shows the build direction. Also, Figure 11 displays collection of surface defects at different locations of the tensile sample built in X orientation. The defects such as porosity and unmelted powder result in lower hardness measurements.

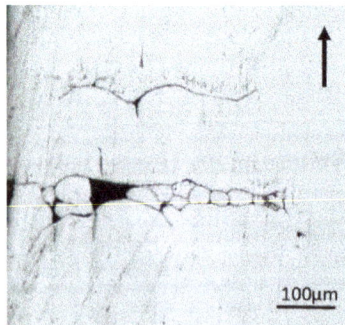

Figure 10. Interlayer defects present in the TZ sample.

Table 6. Average HRC comparison on bottom and top layers of the tensile samples.

Build Orientation	Ave. HRC Bottom Layer	Ave. HRC Top Layer
TX	33.55	31.65
TY	32.77	24.01
TZ	37.67	30.10

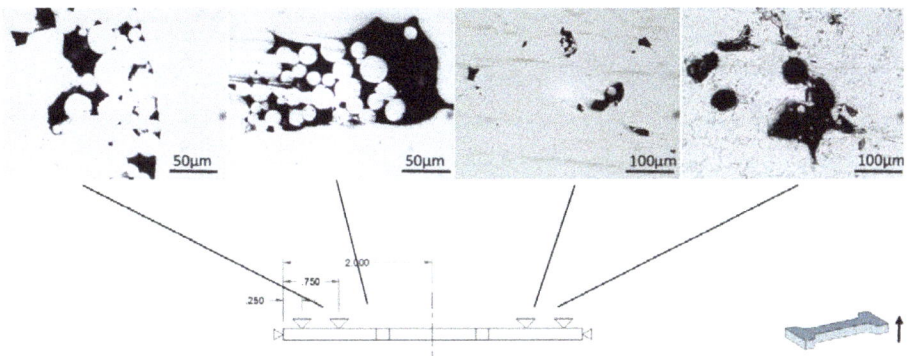

Figure 11. Collection of defects in the TX sample.

3.5. Surface Roughness

One of the characteristics of the SLM process is the rough surface finish. Optical profilometry was used to investigate the surface roughness. The surface properties of as-built fracture toughness samples were studied. Although the results were not all consistent, the side walls exhibited rougher surfaces compared to the top surfaces. A common surface roughness parameter is R_a which is the arithmetic mean of absolute value for linear profiling. The R_a value was about 8 μm at the top surface, while it was about 18 μm at the lateral surfaces. It should be noted that lateral surfaces represent multiple additively manufactured layers, while top surfaces represent a single layer with multiple raster lines. Figure 12 shows the surface roughness map and SEM micrograph for the lateral surface of the TZ sample which represents a relatively rough surface. The crack line in Figure 12 formed due to insufficient interlayer bonding perpendicular to the build orientation. The areas of insufficient fusion can be identified by valleys (dark areas) in the optical surface profile presented in Figures 12 and 13. Figure 13 shows the optical surface profile and SEM micrograph for the top surface of the TX sample. The top surface is clearly smoother than the side surfaces.

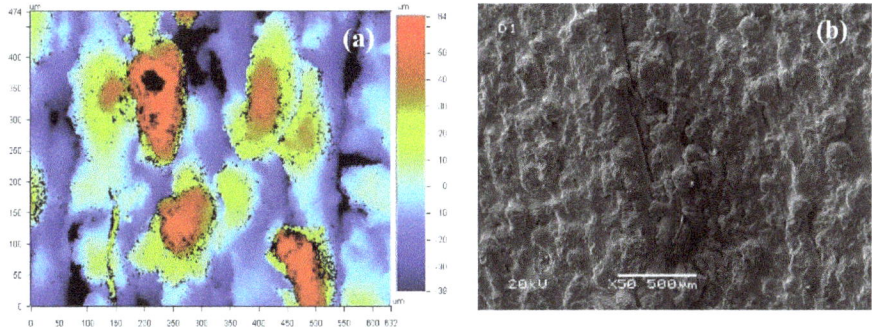

Figure 12. Optical surface profile: (a) and SEM micrograph (b) for the lateral surface of TZ sample.

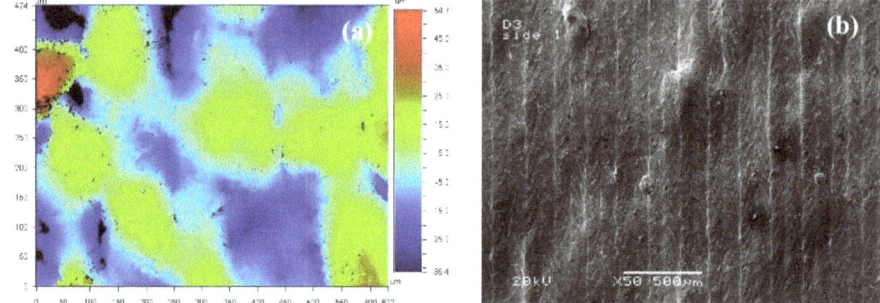

Figure 13. Optical surface profile: (**a**) and SEM micrograph (**b**) for the top surface of TX sample.

EONSINT M270 DMLS generated samples by Chauke et al. [27] exhibited a similar surface morphology. Their samples also showed unmelted and half-melted powder globules, as evident in Figure 14 [27]. The inability of layers to completely join might be due to the absence of sufficient power exposure to melt the previously deposited layers. Cracks and lack of fusion were observable in the remaining samples with different orientations as well.

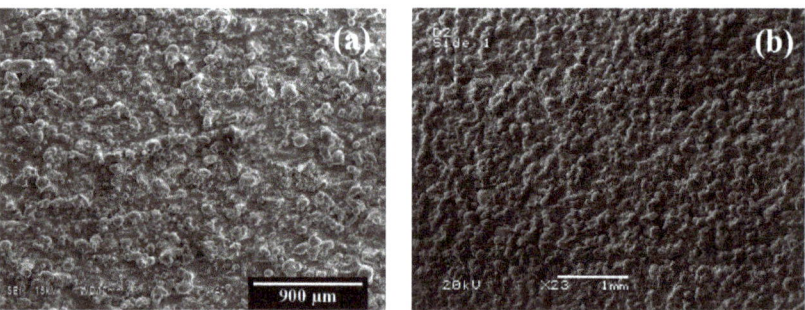

Figure 14. Surface finish of (**a**) EONSINT M270 sample [27]; (**b**) AM250 sample (present study).

Todd et al. [35] measured the surface roughness of a number of samples using optical profilometry. The reported values varied between 5 and 15 µm [35]. The surface roughness values in the present study ranged from 6 to 20 µm, but the average surface roughness was comparable with other reported results, as shown in Table 7.

Table 7. Surface roughness comparison of the present study with other reported data [35].

Ti–6Al–4V	Treatment	R_a (µm)
Wrought [35]	Machined, polished	1
DMLS [35]	As fabricated	11–13
DMLS [35]	Polished	10
DMLS [35]	Electro polished	13
SLM (Present Study)	As fabricated	6–20

4. Conclusions

In this research, the effect of build orientation on the microstructure, mechanical and surface properties of selective laser-melted Ti–6Al–4V alloy was studied. The as-built surfaces along with fracture surfaces were studied by SEM. The tensile test revealed lower yield and ultimate tensile strength in the samples printed in the Z orientation with brittle planar fracture features perpendicular to the build direction. Tensile samples built in the X and Y orientations exhibited brittle fracture features in areas close to the substrate and ductile fracture features in the area farther from the substrate. The SEM images of as-built and fracture surfaces revealed defects including pores, cracks, and unmelted/partially-melted powder particles. The mechanical test results were clearly influenced by the defects. However, the tensile and fracture toughness test results were consistent with previously reported data. Metallography showed identical α + β microstructure in all build orientations, as the samples were all heat treated. The bottom layers close to the substrate showed lower hardness values as compared to the top layers, due to the change in cooling rate. Defects revealed from metallography observations on the surfaces influenced the hardness results. The surface analysis showed variable roughness data on different surfaces. While lateral surfaces showed higher roughness values, the top surface exhibited smoother features. Despite orientation factors and build defects, the surface roughness values were comparable with other studies. It was concluded that microstructure and properties were affected by build orientation in Ti–6Al–4V alloys processed by SLM. Defects influenced the results significantly. Optimization of process parameters may improve the overall quality of AM samples and provide more uniform properties in different build orientations.

Author Contributions: M.E. conceived and designed the experiments; P.H. performed the experiments; P.H. and M.E. analyzed the data and wrote the paper.

Funding: This research received no external funding.

Acknowledgments: The authors would like to thank Behzad Bavarian from California State University, Northridge for his assistance with mechanical testing and scanning electron microscopy. The authors also acknowledge California State University, Los Angeles and National Science Foundation through Grant PREM-1523588 for their support.

Conflicts of Interest: The Authors declare no conflict of interest.

References

1. Gong, H.; Rafi, K.; Gu, H.; Starr, T.; Stucker, B. Analysis of defect generation in Ti-6Al-4V parts made using powder bed fusion additive manufacturing processes. *Addit. Manuf.* **2014**, *1*, 87–98. [CrossRef]
2. *Introduction to Additive Manufacturing Technology*; European Powder Metallurgy Association: Shrewsbury, UK, 2013; p. 6.
3. Rafi, H.K.; Karthik, N.V.; Gong, H.; Starr, T.L.; Stucker, B.E. Microstructures and mechanical properties of Ti6Al4V parts fabricated by selective laser melting and electron beam melting. *J. Mater. Eng. Perform.* **2013**, *22*, 3872–3883. [CrossRef]
4. Sieniawski, J.; Ziaja, W.; Kubiak, K.; Motyka, M. Microstructure and Mechanical Properties of High Strength Two-Phase Titanium Alloys. In *Titanium Alloys-Advances in Properties Control*; Sieniawski, J., Ziaja, W., Eds.; InTech: Gdańsk, Poland, 2013; pp. 69–80.
5. Simonelli, M.; Tse, Y.Y.; Tuck, C. The formation of α + β microstructure in as-fabricated selective laser melting of Ti-6Al-4V. *J. Mater. Res.* **2014**, *29*, 2028–2035. [CrossRef]
6. Barriobero-Vila, P.; Gussone, J.; Haubrich, J.; Sandlöbes, S.; da Silva, J.C.; Cloetens, P.; Schell, N.; Requena, G. Inducing stable α + β microstructures during selective laser melting of Ti-6Al-4V using intensified intrinsic heat treatments. *Materials* **2017**, *10*, 268. [CrossRef] [PubMed]
7. Dadbakhsh, B.V.J.-P.K.S.; Vleugels, J.V.H.K.K.J. Additive Manufacturing of Metals via Selective Laser Melting: Process Aspects and Material Developments. In *Additive Manufacturing*; Sudarshan, T.S.S.T.S., Ed.; CRC Group, Taylor & Francis Group: Boca Raton, NJ, USA, 2015; pp. 69–99.
8. Kruth, J.P.; Levy, G.; Klocke, F.; Childs, T.H.C. Consolidation phenomena in laser and powder-bed based layered manufacturing. *CIRP Ann. Manuf. Technol.* **2007**, *56*, 730–759. [CrossRef]
9. Childs, T. Selective laser sintering (melting) of stainless and tool steel powders: Experiments and modelling. *Proc. Inst. Mech. Eng. Part B* **2005**, *219*, 339–357. [CrossRef]
10. Das, S. Physical Aspects of Process Control in Selective Laser Sintering of Metals. *Adv. Eng. Mater.* **2003**, *5*, 701–711. [CrossRef]
11. Fatemi, S.A.; Ashany, J.Z.; Aghchai, A.J.; Abolghasemi, A. Experimental investigation of process parameters on layer thickness and density in direct metal laser sintering: A response surface methodology approach. *Virtual Phys. Prototyp.* **2017**, *12*, 133–140. [CrossRef]

12. Mohanty, P.S.; Mazumder, J. Solidification behavior and microstructural evolution during laser beam—Material interaction. *Metall. Mater. Trans. B* **1998**, *29*, 1269–1279. [CrossRef]
13. Lavernia, E.J.; Srivatsan, T.S. The rapid solidification processing of materials: Science, principles, technology, advances, and applications. *J. Mater. Sci.* **2010**, *45*, 287–325. [CrossRef]
14. Birol, Y. Microstructural characterization of a rapidly-solidified Al-12 wt% Si alloy. *J. Mater. Sci.* **1996**, *31*, 2139–2143. [CrossRef]
15. Sing, S.L.; An, J.; Yeong, W.Y.; Wiria, F.E. Laser and electron-beam powder-bed additive manufacturing of metallic implants: A review on processes, materials and designs. *J. Orthop. Res.* **2016**, *34*, 369–385. [CrossRef] [PubMed]
16. Lewandowski, J.J.; Seifi, M. Metal Additive Manufacturing: A Review of Mechanical Properties. *Annu. Rev. Mater. Res.* **2016**, *46*, 151–186. [CrossRef]
17. Sames, W.J.; List, F.A.; Pannala, S.; Dehoff, R.R.; Babu, S.S. The metallurgy and processing science of metal additive manufacturing. *Int. Mater. Rev.* **2016**, *61*, 315–360. [CrossRef]
18. Sing, S.L.; Wiria, F.E.; Yeong, W.Y. Selective laser melting of titanium alloy with 50wt% tantalum: Effect of laser process parameters on part quality. *Int. J. Refract. Met. Hard Mater.* **2018**, *77*, 120–127. [CrossRef]
19. Simonelli, M.; Tse, Y.Y.; Tuck, C. Effect of the build orientation on the mechanical properties and fracture modes of SLM Ti-6Al-4V. *Mater. Sci. Eng. A* **2014**, *616*, 1–11. [CrossRef]
20. Edwards, P.; Ramulu, M. Effect of build direction on the fracture toughness and fatigue crack growth in selective laser melted Ti-6Al-4-‰V. *Fatigue Fract. Eng. Mater. Struct.* **2015**, *38*, 1228–1236. [CrossRef]
21. Barriobero-Vila, P.; Gussone, J.; Stark, A.; Schell, N.; Haubrich, J.; Requena, G. Peritectic titanium alloys for 3D printing. *Nat. Commun.* **2018**, *9*, 3426. [CrossRef] [PubMed]
22. *Ti6Al4V ELI-0406 Powder for Additive Manufacturing*; Renishaw plc: Staffordshire, UK, 2017.
23. ASTM. E8/E8M standard test methods for tension testing of metallic materials 1. In *Annu. B. ASTM Stand. 4*; ASTM International: West Conshohocken, PA, USA, 2010; Volume I, pp. 1–27.
24. E08-Committee. *ASTM E399-97—Standard Test Method for Plane-Strain Fracture Toughness of Metallic Materials 1*; ASTM International: West Conshohocken, PA, USA, 1997; Volume E399–90.
25. Yuan, P.; Gu, D.; Dai, D. Particulate migration behavior and its mechanism during selective laser melting of TiC reinforced Al matrix nanocomposites. *Mater. Des.* **2015**, *82*, 46–55. [CrossRef]
26. Seifi, M.; Dahar, M.; Aman, R.; Harrysson, O.; Beuth, J.; Lewandowski, J.J. Evaluation of Orientation Dependence of Fracture Toughness and Fatigue Crack Propagation Behavior of As-Deposited ARCAM EBM Ti-6Al-4V. *JOM* **2015**, *67*, 597–607. [CrossRef]
27. Chauke, L.; Mutombo, K.; Kgomo, C. Characterization of the direct metal laser sintered Ti6Al4V Components. In Proceedings of the RAPDASA 2013 Conference, Clarens, South Africa, 29 October–1 November 2013.
28. ASTM International. *BS ISO/ASTM 52900:2015 Additive Manufacturing. General Principles. Terminology*; Rapid Manufacturing Association: West Conshohocken, PA, USA, 2013; pp. 10–12.
29. Herderick, E. Additive manufacturing of metals: A review. *Mater. Sci. Technol. Conf. Exhib.* **2011**, *2*, 1413–1425.
30. Cain, V.; Thijs, L.; van Humbeeck, J.; van Hooreweder, B.; Knutsen, R. Crack propagation and fracture toughness of Ti6Al4V alloy produced by selective laser melting. *Addit. Manuf.* **2015**, *5*, 68–76. [CrossRef]
31. Van Hooreweder, B.; Moens, D.; Boonen, R.; Kruth, J.-P.; Sas, P. Analysis of Fracture Toughness and Crack Propagation of Ti6Al4V Produced by Selective Laser Melting. *Adv. Eng. Mater.* **2012**, *14*, 92–97. [CrossRef]
32. Becker, T.H.; Beck, M.; Scheffer, C. Microstructure and mechanical properties of Direct Metal Laser Sintered Ti-6Al-4V. *S. Afr. J. Ind. Eng.* **2015**, *26*, 1–10. [CrossRef]
33. Vrancken, B.; Thijs, L.; Kruth, J.-P.; van Humbeeck, J. Heat treatment of Ti6Al4V produced by Selective Laser Melting: Microstructure and mechanical properties. *J. Alloys Compd.* **2012**, *541*, 177–185. [CrossRef]
34. Poondla, N.; Srivatsan, T.S.; Patnaik, A.; Petraroli, M. A study of the microstructure and hardness of two titanium alloys: Commercially pure and Ti-6Al-4V. *J. Alloys Compd.* **2009**, *486*, 162–167. [CrossRef]
35. Mower, T.M.; Long, M.J. Mechanical behavior of additive manufactured, powder-bed laser-fused materials. *Mater. Sci. Eng. A* **2016**, *651*, 198–213. [CrossRef]

© 2018 by the authors. Licensee MDPI, Basel, Switzerland. This article is an open access article distributed under the terms and conditions of the Creative Commons Attribution (CC BY) license (http://creativecommons.org/licenses/by/4.0/).

Article

Effect of SLM Build Parameters on the Compressive Properties of 304L Stainless Steel

Okanmisope Fashanu [1], Mario F. Buchely [2], Myranda Spratt [2], Joseph Newkirk [2], K. Chandrashekhara [1,*], Heath Misak [3] and Michael Walker [3]

1. Department of Mechanical and Aerospace Engineering, Missouri University of Science and Technology, Rolla, MO 65409, USA; oaff4x@mst.edu
2. Department of Materials Science and Engineering, Missouri University of Science and Technology, Rolla, MO 65409, USA; buchelym@mst.edu (M.B.); msf998@mst.edu (M.S.); jnewkirk@mst.edu (J.N.)
3. Spirit AeroSystems, Wichita, KS 67210 USA; heath.e.misak@spiritaero.com (H.M.); michael.a.walker@spiritaero.com (M.W.)
* Correspondence: chandra@mst.edu

Received: 23 April 2019; Accepted: 29 May 2019; Published: 2 June 2019

Abstract: Selective laser melting (SLM) is well suited for the efficient manufacturing of complex structures because of its manufacturing methodology. The optimized process parameters for each alloy has been a cause for debate in recent years. In this study, the hatch angle and build orientation were investigated. 304L stainless steel samples were manufactured using three hatch angles (0°, 67°, and 105°) in three build orientations (x-, y-, and z-direction) and tested in compression. Analysis of variance and Tukey's test were used to evaluate the obtained results. Results showed that the measured compressive yield strength and plastic flow stress varied when the hatch angle and build orientation changed. Samples built in the y-direction exhibited the highest yield strength irrespective of the hatch angle; although, samples manufactured using a hatch angle of 0° exhibited the lowest yield strength. Samples manufactured with a hatch angle of 0° flowed at the lowest stress at 35% plastic strain. Samples manufactured with hatch angles of 67° and 105° flowed at statistically the same flow stress at 35% plastic strain. However, samples manufactured with a 67° hatch angle deformed non-uniformly. Therefore, it can be concluded that 304L stainless steel parts manufactured using a hatch angle of 105° in the y-direction exhibited the best overall compressive behavior.

Keywords: selective laser melting (SLM); compression testing; stainless steel; hatch angle; build orientation; analysis of variance; Tukey's test

1. Introduction

The demand for stronger, lighter, and more customizable parts has driven the development and research of new manufacturing methods, tools, and technologies. In this sense, the development and continuous improvement of manufacturing methods have dramatically changed the way designers and engineers pursue design and manufacturing [1]. Selective laser melting (SLM), a powder-bed fusion process of metal additive manufacturing (AM), involves the production of dense parts from a 3D computer-aided design model by the selective melting of metal powder by using a laser heat source. The SLM process is a timely and cost-effective method of building complex geometries that are impossible to manufacture using conventional processes [2].

During part fabrication in SLM, fine metal powder is introduced into the build chamber by a feeding system or powder hopper, and a soft distribution recoater blade is used to drag the powder across the build plate. A high-powered laser is then used to selectively melt the powder together to form a finished part based on the principles of rapid prototyping [3]. The complexity of the SLM process makes it difficult to characterize and understand the mechanical performance of parts made using this technique [4,5]. In AM, part anisotropy and mechanical performance are strongly affected by the process parameters. By varying the process parameters, the mechanical properties can be optimized. Many investigators have studied the effects of process parameters on the behavior of additively manufactured parts. For example, Popovich et al. [6] investigated the anisotropy of mechanical properties of parts manufactured using SLM. They found a dependence of the mechanical properties of Ti-6Al-4V on the build orientation. From their study, it was found that the strength of the produced part is dependent on the grain growth direction, which is controlled by the build orientation. Miranda et al. [7] developed a predictive model for the physical and mechanical properties of 316L stainless steel. They observed changes in the mechanical properties of the steel when the laser speed, scanning speed, and scanning spacing was changed. They attributed these changes in mechanical properties to variations in densification levels and residual porosity. The effects of build size, build orientation, and part thickness on the tensile properties of 304L stainless steel has also been studied by Ortiz Rios et al. [8]. During their study, they observed that the part size had no effects on the mechanical properties, however, part orientation did. Guan et al. [3] and Anam et al. [9] individually tried to investigate the hatch angle used during the SLM process. They both investigated different hatch angle sets and used different methods of assessments in their studies. Guan et al. concluded that a hatch angle of 105° produced the best part with respect to tensile strength while Anam et al. concluded that a 67° hatch angle produced the best part with respect to microstructure. Other works available in literature with respect to the effects of process parameters on the mechanical properties of AM parts can be found in [10–12].

304L stainless steel, a type of authentic steel, has gained a lot of interest over the years due to its chemical composition and mechanical properties [13]. When 304L stainless steel is used for part production in SLM, the low carbon content minimizes deleterious carbide precipitation, which minimizes the need for solution annealing. Some of the available works on SLM of 304L stainless steel can be found in [8,13–16]. SLM manufactured 304L stainless steel exhibits higher mechanical strength (yield and ultimate tensile strength) over conventionally manufactured 304L stainless steel, which makes it applicable for use in salt-water body applications that require high strength.

At the completion of an in-depth literature review, it was observed that although many works exist with respect to SLM process parameters, there is not enough information about the hatch angle and how it affects the mechanical properties of manufactured parts. The majority of available works considered a hatch angle of 67° in their studies; however, Guam et al. [3] claimed 105° produced better parts. It was also observed that most of the available process parameter investigations only considered tensile stress–strain curves, which in some cases does not represent the complete behavior of a material. In this work, the mechanical performance of SLM 304L stainless steel was investigated with respect to changes in the hatch angle and build orientation. Test specimens were built with three hatch angles (0°, 67°, and 105°) in three build orientations (x, y, and z) and tested in quasi-static compression. Build orientation was considered because previous works show that build orientation affects the mechanical properties of the SLM parts [8]. The yield strength and plastic flow stress at 35% plastic strain were evaluated. A two-way analysis of variance (two-way ANOVA) technique and Tukey's test were used to evaluate the difference in mechanical responses caused by changes in the hatch angles and build orientations, while also considering hatch angle–build orientation interaction. The ANOVA technique is the most commonly used statistical tool for investigating effects and interactions between two or more factors. Some available studies using ANOVA in AM can be found in [17,18].

2. Materials and Methods

2.1. Fabrication

Argon gas atomized 304L stainless steel powder (Figure 1), ranging in particle size between 15 μm and 63 μm, was purchased from LPW technology and used in this study. The chemical composition of the powder is shown in Table 1. Kriewall et al. [13] conducted a detailed investigation on the powder used in this work. Octagonal samples were manufactured in an argon-filled environment using a Renishaw AM250 SLM machine (machine parameters are summarized in Table 2). Octagonal geometry was selected due to the convenience for machining samples built in the x- and y-direction.

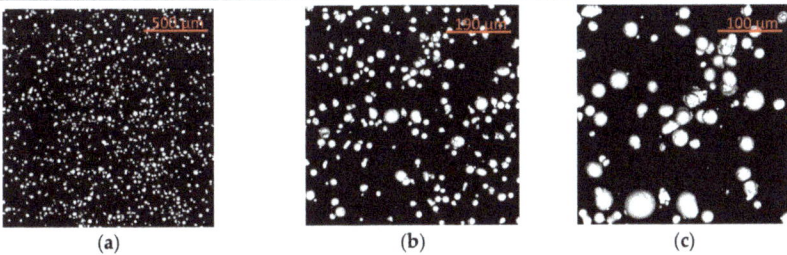

Figure 1. SEM observations of the 304L powder used in this work, showing its morphology at different magnifications: (**a**) 500 µm, (**b**) 190 µm, and (**c**) 100 µm.

Table 1. Chemical composition (in wt.%) of 304L stainless steel powder.

Element	Cr	Ni	Mn	Si	Cu	N	O	C	P	S	Fe
wt.%	18.5	9.9	1.4	0.63	0.1	0.09	0.02	0.015	0.012	0.004	Balance

Table 2. Selective laser melting (SLM) machine parameters.

Parameter	Value
Laser type	1070 nm NdYAG
Laser power (W)	200
Effective build volume (mm × mm × mm)	248 × 248 × 280
Laser spot	70
Hatch distance (mm)	0.085
Fill pattern	Stripes
Exposure time (µs)	75
Point distance (µm)	60
Layer thickness (µm)	50
Inert gas during production	Argon

In SLM, the selected scan strategy controls the shape of the melt pool and the resulting microstructure. Hatch angle at θ° can be defined as the angle between the scanning directions of two immediate scan layers, as shown in Figure 2. The hatch angle controls the variance in the 360° space, the spacing between similarly oriented layers, and beam titling. As there are 360 possible scanning directions, there are 360 possible hatch angles. Hatch angles at 67° and 105° were selected for this study because they have been studied by other investigators and are known to produce parts with excellent properties [3,9], while a 0° hatch angle was selected to investigate the effects of no rotation between consecutive layers.

Octagonal samples with a side length of 3.84 mm and height of 27.80 mm were manufactured with three different hatch angles and in three distinct orientations (Table 3), subsequently referred to as configurations (a) to (i). The x-direction (0°) was taken as the reference orientation (see Figure 3). The other two build directions (y-direction and z-direction) were obtained by rotating the reference sample (x-direction) 90° around the z- and y-axis, respectively. The direction of the height (longest side) of the octagonal cylinder was in correlation with the build direction in accordance to ISO/ASTM 52921 standard [19,20]. These three build directions (x-, y-, and z-direction) were considered for investigation because structures built using these orientations require little or no support material. After manufacturing, compression specimens were prepared for compression testing by machining the octagons into solid cylinders (diameter 6.35 ± 0.07 mm, height 6.35 ± 0.30 mm) using a computer numerical controlled lathe (at 250 rpm spindle and 0.006 in/rev feed). The samples were machined to produce smooth surfaces and flat parallel ends required for accurate testing.

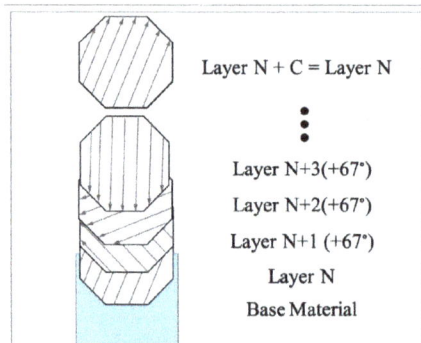

Figure 2. Schematic view of the scan direction in-between layers (configuration (f) in Table 3).

Table 3. Summary of different build configurations studied during this study.

Configuration	Hatch Angle (°)	Build Orientation
(a)	0	x
(b)	0	y
(c)	0	z
(d)	67	x
(e)	67	y
(f)	67	z
(g)	105	x
(h)	105	y
(i)	105	z

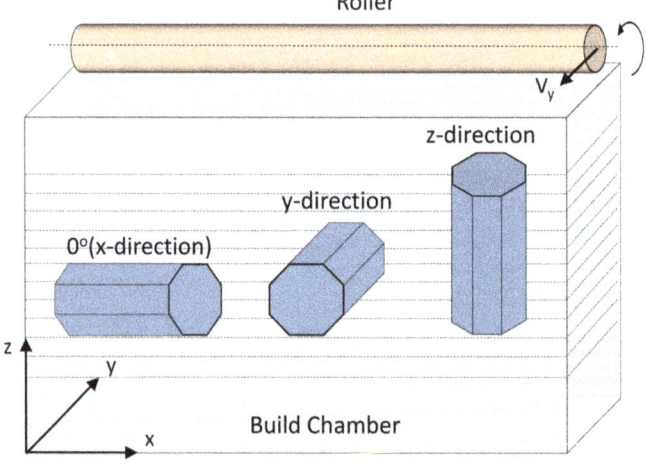

Figure 3. Schematic view of build orientations for manufacturing SLM parts.

2.2. Experimental Testing

2.2.1. Porosity

The percentage porosity in the manufactured parts was measured using the Archimedes' method (ASTM B962). The samples were assumed to have little to no surface connected porosity, so the saturated weight was not measured, and oil impregnation of the parts was not done. The machined samples' dry weight and suspended weight were used to calculate the bulk density of the parts (Equation (1)) while the ratio of the measured density and the density of 304L stainless steel was used to calculate the percentage porosity of the manufactured parts (Equation (2)).

$$\rho^* = D/(D - S) \tag{1}$$

$$\% \text{ porosity} = 1 - (\rho^*/\rho \times 100) \tag{2}$$

where ρ^* is the measured density, D is the dry weight of the specimen, S is the suspended weight, and ρ is the bulk density of stainless steel taken as 8.00 g/cc [21].

2.2.2. Compression Tests

Compression tests for each experimental case were performed using an MTS 380 frame, according to ASTM E9-09 standard [22]. Three samples were compressed per each case to check repeatability of the data. The crosshead speed of the frame was fixed to obtain an initial strain rate of 5×10^{-3} min^{-1} in the sample. Force and displacement changes were tracked during tests and used to plot the stress–strain curve. The machine crosshead displacement and load were converted into true stress–true strain using the following equations (Equation (3a–d)):

$$\sigma = (F/A_0) \tag{3a}$$

$$\varepsilon = (L - L_0)/L_0 \tag{3b}$$

$$\sigma_T = \sigma(1 - \varepsilon) \tag{3c}$$

$$\varepsilon_T = \ln(1 - \varepsilon) \tag{3d}$$

where F is the measured force (N), A_0 is the cross-sectional area of the sample (m^2), L_0 is the initial length of the sample (m), L is the final length of the sample (m), σ is the engineering stress (Pa), ε is the engineering strain (m/m), σ_T is the true stress (Pa), and ε_T is the true strain (m/m).

In order to compare and analyze the experimental data, two data points (yield strength and flow stress at 35% plastic strain) on the stress–strain curve were selected. The yield strength was selected because of its importance in part design. 35% plastic strain was selected because it was observed during testing that at this point, samples built with hatch angles of 0° and 67° showed profound non-uniform deformation, inferring that the engineering to true stress conversion was not valid at strains higher than this value. This non-uniform deformation will be presented later in this paper.

2.2.3. Examination

Micrographs of the machined and untested samples were taken using an optical microscope. The machined samples were mounted in Bakelite. They were then ground using 320 SiC paper to the desired area. The samples were polished with diamond solution to 1 micron with a final polish of 0.05 micron colloidal silica. A 60:40 nitric acid:water electrolyte ratio was used to facilitate electrolytic polishing, which was done at 6 V for 10 s. The machined samples were cut and prepared according to ASTM E3-11 [23].

Due to the observance of non-uniform deformation after compression, surface aspect ratio measurement and calculations were carried out in the tested samples. Equation (4) was used to calculate the surface aspect ratio. The input parameters used in Equation (4) were obtained by measurements of the longest and shortest Ferets of each compressed sample using image J. Figure 4 shows the visual representation of the measured parameters where M_f is the length of the major Feret (m) and N_f is length of the minor Feret (m).

$$A_s = (M_f/N_f) \tag{4}$$

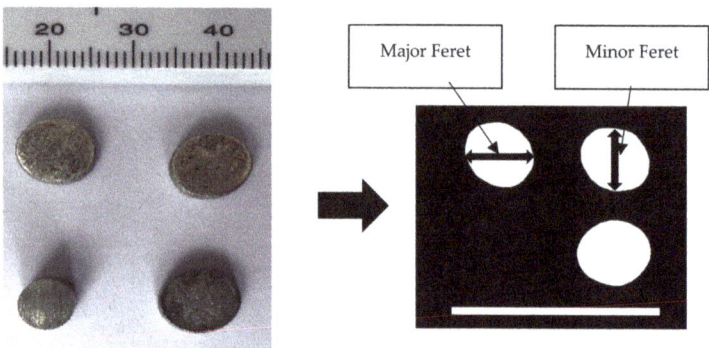

Figure 4. Major and minor Feret measurement illustration.

2.3. Statistical Analysis

In this study, two-way ANOVA was used in the analysis of the compression data. In a two-way ANOVA, the means of two groups of independent factors are compared. The aim of a two-way ANOVA is to test if there exists an interaction between the two independent variables on the dependent variable [24]. In a two-way ANOVA, the interaction term checks if the effect of one independent variables on the dependent variable is equal for all values of the other independent variable. The effects of hatch angle and build orientation were evaluated using a two-way ANOVA. These two build parameters, hatch angle and build orientation, were the independent variables (factors). The response variables were the yield strength and plastic flow stress at 35% plastic strain. These two points on the stress–strain curve (yield strength and flow stress at 35% plastic strain) were selected to investigate the effects of hatch angle and build orientation on the elastic and plastic properties of SLM 304L stainless steel. Table 4 shows the full-factorial design, while Table 5 shows the breakdown of the ANOVA table used during this analysis. For each case, three replicates were tested. The mathematical model used during this analysis is as follows [24]:

$$Y_{ijk} = \mu + \alpha_i + \beta_j + (\alpha\beta)_{ij} + \epsilon_{ijk}, \epsilon \sim iid\ N(0, \sigma^2) \tag{5}$$

where Y_{ijk} is the kth observation at the ith factor A level and jth factor B level, μ is the overall mean, α_i is the factor effect of factor A at level i, β_j is the factor effect of factor B at level j, $(\alpha\beta)_{ij}$ represents the interaction effect of factors A and B, ϵ_{ijk} is the random error, and $\epsilon \sim iid\ N(0, \sigma^2)$ is a restriction placed on the error term, meaning the error terms are independent and identically distributed. These error terms are distributed normally around a zero mean value and a variance of 'σ^2'.

Table 4. Full-factorial design.

Factors	Levels		
Hatch angle (I)	0°	67°	105°
Build orientation (II)	x-direction	y-direction	z-direction

Table 5. ANOVA table for the factorial experiment.

Source	Degrees of Freedom (DF)	Sum of Squares (SS)	Mean Square (MS)	F Ratio
Treatment combinations	$a*b - 1$	$SS_{Treat.comb}$	$MS_{Treat.\ comb}$	$MS_{Treat.\ Comb}/MS_{Error}$
Hatch angle (I)	$a - 1$	SS_I	MS_I	MS_I/MS_{Error}
Build orientation (II)	$b - 1$	SS_{II}	MS_{II}	MS_{II}/MS_{Error}
Factor I*II	$(a - 1)*(b - 1)$	SS_{I*II}	MS_{I*II}	MS_{I*II}/MS_{Error}
Error	$a*b*(n - 1)$	SS_{Error}	MS_{Error}	
Total	$(a*b*n) - 1$	SS_{Total}		

$SS_{Treat.Comb}$, SS_I, SS_{II}, and SS_{I*II} are the sum of squares due to deviations from $H0_{\mu ij}$, $H0_I$, $H0_{II}$, and $H0_{I*II}$ respectively, a is the number of levels of hatch angle, b is the number of levels of build orientation and n is the number of replications.

3. Results and Discussions

3.1. Microstructural Analysis and Relative Density

The microstructure of the samples was inspected in the as-built state. Some of the optical images obtained are shown in Figure 5. The microstructure is typical of SLM-printed 304L etched with an electrolytic etchant [25,26], and consists of nearly 100% austenitic with a small amount of delta-ferrite phase, as reported by Amine et al. [25] for the same SLM 304L stainless steel. From Figure 5, the melt pool boundary can be seen as a thin white line, either in a cup shape or in a relatively straight line (depending on the orientation of the mounted specimen). The interior of the melt pool includes bright and dark regions, which consists of a cellular structure. The cellular structure is a fine feature that can be better visualized by higher resolution imaging, as shown in Figure 6 [27]. In 316L, which shows a similar microstructure to 304L, the cellular structure was found to contain highly entangled dislocations, and to be associated to the segregation of Cr and Mo to the cell walls [28]. It is likely that the electrolytic etchant used in this study preferentially etched the cell walls due to the concentration of dislocations there, which led to their distinct visibility in the optical microscope.

Figure 5. Microstructure side-by-side comparison of (a) 0°, (b) 67°, and (c) 105° hatch angle specimens built in the y-direction before compression test. Electrolytic etchant in 60:40 nitric acid:water solution.

The density measurements done by Archimedes' method resulted in densities ranging from 98.6% to 98.8% dense. There were no clear trends in the data, indicating that the hatch angle rotation and build orientation did not noticeably affect the density of the manufactured parts. Defects in SLM processes do not tend to be random due to the layer-by-layer nature of the process. This makes microstructural evaluation of defect distribution and volume difficult, as it is unknown if the microstructural image adequately captures these periodic defects. For this reason, a quantitative measurement of porosity via microstructural evaluation was not performed. Qualitatively, there were also no obvious differences in the microstructural porosity between specimens. Several pores can be seen in the images in Figure 5, confirming that full density was not achieved.

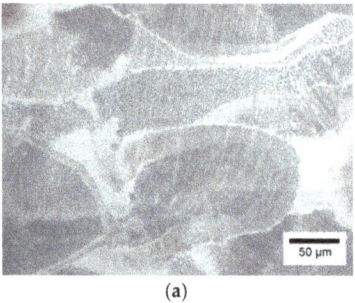 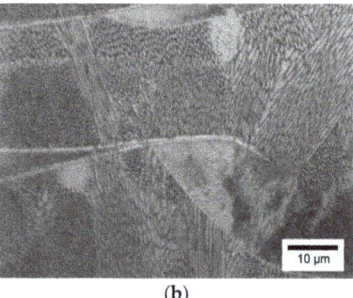

(a)   (b)

Figure 6. Close inspection of the 0° hatch angle, y-direction specimen showing the cellular structure and melt pool boundary, at two different magnifications: (**a**) 50 µm and (**b**) 10 µm.

3.2. Compressive Behavior

After compression, the stress–strain curves were developed using Equations (3) and (4). The strains of these curves were calculated using the machine displacement. Due to the specimen size, inaccurate strain measurements were obtained at low displacements; therefore, the elastic moduli could not be evaluated. However, it was possible to compare the slopes of the different stress–strain curves. Figure 7 shows two examples of such comparisons.

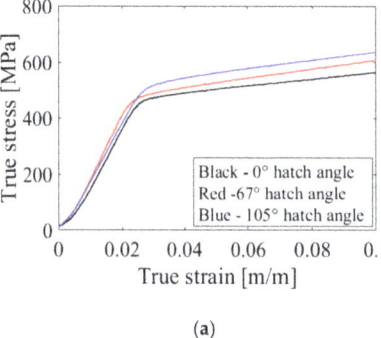 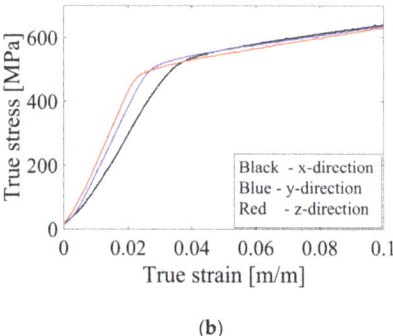

(a)   (b)

Figure 7. True stress–strain curve of SLM 304L stainless steel showing the elastic region: (**a**) samples built with three hatch angles in the y-direction, (**b**) samples built in three orientations using hatch angle of 105°.

From Figure 7, it can be seen that by varying the hatch angle and build orientation, little changes (less than 10% in both cases) can be observed in the slopes of the stress–strain curves. It was assumed that these little variations were as a result of different experimental conditions.

After yielding, non-uniform plastic deformation, rather than barreling at the midpoint, was observed in some samples at high strains. This non-uniform plastic deformation has never been observed or reported in literature. Figure 8 shows the final geometry of the samples after compression. Samples built with a 0° hatch in the z-direction sheared and ovalled (Figure 8c), while samples built in the same orientation with a 67° hatch angle only sheared (Figure 8f). It was also observed that samples built with hatch angles of 0° and 67° formed an ellipsoid when built in the x-direction (Figure 8a and 8d). However, only samples built with a hatch angle of 0° sheared and formed an ellipsoid when built in the y-direction (Figure 8b). No shearing or ovalling was observed in samples built using a hatch angle of 105° after compression (Figure 8g–i).

The degree of non-uniformity was evaluated quantitatively by surface aspect ratio calculations using Equation (4). Figure 9 shows the average surface aspect ratios calculated. From Figure 9, it can be observed that samples built with a 0° hatch angle were the least circular after compression followed by samples built using a 67° and 105° hatch angle, respectively. Samples built with a hatch angle of 105° had similar circularities in all build orientations.

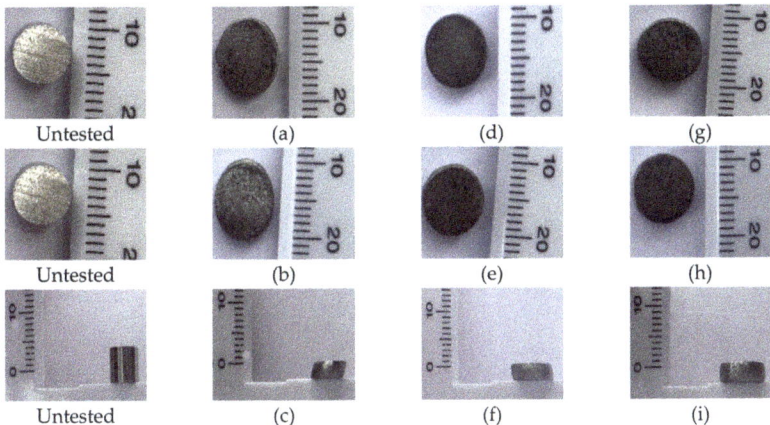

Figure 8. Geometry of untested and tested samples (where **a–c** equals samples built with a 0° hatch angle in the x-, y-, and z-direction, respectively, **d–f** equals samples built with a 67° hatch angle in the x-, y-, and z-direction, respectively, and **g–i** equals samples built with a 105° hatch angle in the x-, y-, and z-direction, respectively).

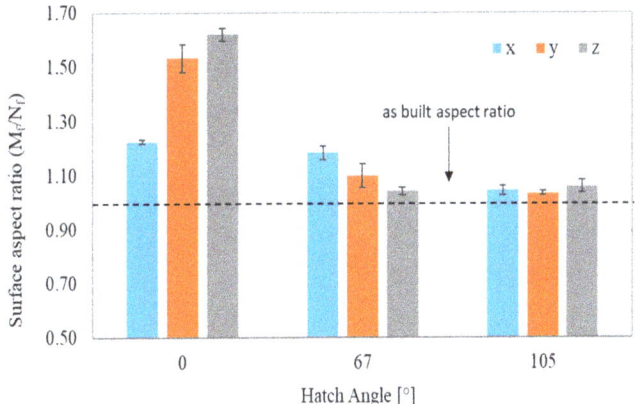

Figure 9. Surface aspect ratio comparison of compressed samples built using different hatch angles and build orientations.

The compressive true stress–plastic strain curves are shown in Figure 10. From Figure 10, the yield strength (stress at 0% plastic strain) and flow stress at 35% plastic strain were extrapolated using a MATLAB code. Table 6 shows the full factorial design adopted for this study as well as the yield strength, flow stress at 35% plastic strain, the sample mean, and standard deviation. From Table 6, it can be seen that the manufactured samples showed higher strengths when the layer structure was parallel to the direction of the force (i.e., x- and y-direction) when compared to samples in which the structure was perpendicular to the loading direction (z-direction). This behavior was also recorded by Meier et al. [29] and Hitzler et al. [16] in their study. The data in Table 6 was used as input data for the two-way ANOVA analysis.

Table 6. Full factorial design of the two control factors with three replicates.

	Factor		Response: Yield Strength (MPa)					Response: Flow Stress at 35% Plastic Strain Strength (MPa)				
			Replication					Replication				
Sample	A	B	1	2	3	Mean	St.Dev	1	2	3	Mean	St.Dev
1	0	x	431	436	424	430	6	892	900	939	911	25
2	0	y	481	492	471	482	10	869	904	854	876	26
3	0	z	442	415	449	436	18	819	851	841	837	16
4	67	x	504	507	513	508	5	990	988	985	987	2
5	67	y	553	528	504	528	25	991	1000	998	993	6
6	67	z	481	479	463	474	10	996	979	992	989	9
7	105	x	522	526	489	512	20	998	998	983	993	9
8	105	y	483	544	518	515	31	997	987	998	995	5
9	105	z	483	497	492	491	7	1007	1008	996	1004	7

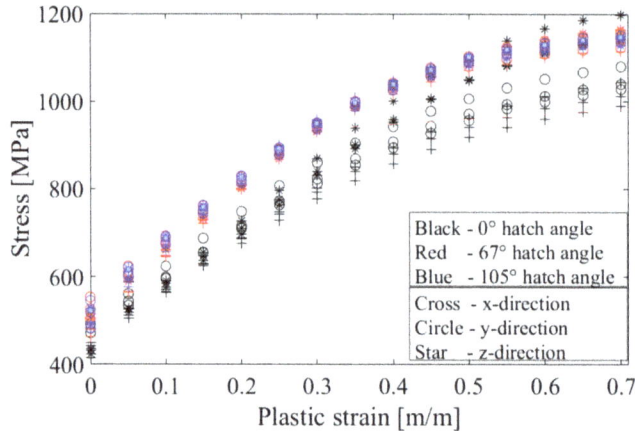

Figure 10. Plastic true stress–strain curve showing the compressive results for all tested conditions.

3.3. Statistical Analysis

A two-way ANOVA was performed using the experimental data in Table 6 to check which hatch angle and build orientation produced the best part with respect to compressive performance. The analysis of variance was carried out at a confidence level of 95% ($\alpha = 0.05$) using JMP 14, a statistical commercial software. It was assumed that the data obtained from the experiments were normally distributed, the variance between the dependent variables were equal and the obtained data were independent and identically distributed. The ANOVA and treatment effects results for the analyzed response variables are reported in Tables 7–10. During this study, four null hypotheses (H_0) concerning the treatment effects were considered. These null hypotheses were considered in order (i.e., 1, 2, ... 4) and are as follows:

1. $H_{0\text{Treat.Comb}}$: $\mu_{11} - \mu_{12} = \ldots = \mu_{ab}$ (tests to see if the treatment combination means are equal). Reject H_0 if $MS_{\text{Treat. Comb}}/MS_{\text{Error}} > F_{\alpha(ab-1,ab(n-1))}$
2. $H_{0I \cdot II}$: $(\alpha\beta)_{ij} = 0$; i,j (tests for the presence of interaction) Rejected $H_{0I \cdot II}$ if $MS_{I \cdot II}/MS_{\text{Error}} > F_{\alpha((a-1)(b-1),ab(n-1))}$
3. H_{0I}: $\alpha_1 = \alpha_2 = \ldots = \alpha_I = 0$ (test to see if there is a difference between the hatch angle means) Reject H_{0I} if $MS_I/MS_{\text{Error}} > F_{\alpha(a-1,ab(n-1))}$
4. H_{0II}: $\beta_1 = \beta_2 = \ldots = \beta_{II} = 0$ (tests to if there is a difference between the build orientation means) Reject H_{0II} if $MS_{II}/MS_{\text{Error}} > F_{\alpha(b-1,ab(n-1))}$

Table 7. ANOVA results for yield strength ($\alpha = 0.05$).

Source	DF	Sum of Squares	Mean Square	F Ratio
Model	8	28,743.027	3592.88	12.4699
Error	18	5186.220	288.12	Prob > F
C. Total	26	33,929.246		<0.0001*

Table 8. Effects test for yield strength ($\alpha = 0.05$).

Source	Nparm	DF	Sum of Squares	F Ratio	Prob > F
Hatch angle	2	2	18,438.828	31.9982	<0.0001*
Build orientation	2	2	7864.314	13.6475	0.0002*
Hatch angle*Build orientation	4	4	2439.885	2.1170	0.1207

Table 9. ANOVA results for flow stress at 35% plastic strain ($\alpha = 0.05$).

Source	DF	Sum of Squares	Mean Square	F Ratio
Model	8	93,496.572	11687.1	58.00850
Error	18	3626.491	201.5	Prob > F
C. Total	26	97,123.062		<0.0001*

Table 10. Effects table for flow stress at 35% plastic strain ($\alpha = 0.05$).

Source	Nparm	DF	Sum of Squares	F Ratio	Prob > F
Hatch angle	2	2	85,115.62	211.2348	<0.0001*
Build orientation	2	2	1897.766	4.7098	0.0226*
Hatch angle*Build orientation	4	4	6483.144	8.0447	0.0007*

Nparm is the number of parameters associated with the effect, DF is the degree of freedom, and '*' means the parameter is significant.

From Tables 7 and 9, it can be observed that the *p*-value for C. total is <0.05, which means there exists a significant difference between the means of the treatment combinations (i.e., the null hypothesis H_0 was rejected and the model can be used to analyze the experimental data). Since the model was found to be significant, the interaction between hatch angle and build orientation was examined using the effects table (Tables 8 and 10) for both response variables.

From Table 8, the interaction effect on the measured yield strength was found to be insignificant (i.e., failed to reject the null hypothesis $H_{0I*II(yield)}$) leading to the investigation of the main effects. The main effects, hatch angle and build orientation, were found to significantly influence the yield strength (i.e., null hypothesis H_{0I} and H_{0II} were rejected). Considering that the main effects were significant, a Tukey's test (Table 11) was conducted on the main effect means to determine which factor levels produced the specimen with highest yield strength. From the Tukey's test and the least squares means plot (Figure 11), it can be deduced that parts built with a hatch angle of 67° and a hatch angle of 105° produced parts with similar yield strengths, while samples built with a 0° hatch angle exhibited lower yield strengths. It can also be seen that parts built in the y-direction exhibited the highest mean average yield strength.

Table 11. Least square means (LS Means) Differences Tukey honestly significant difference (HSD) (yield strength); $\alpha = 0.050$.

Level		Least Sq. Mean
105	A	506.07706
67	A	503.27145
0	B	449.29164
y	A	508.34234
x	B	483.49386
z	B	466.80394

Levels not connected by same letter are significantly different.

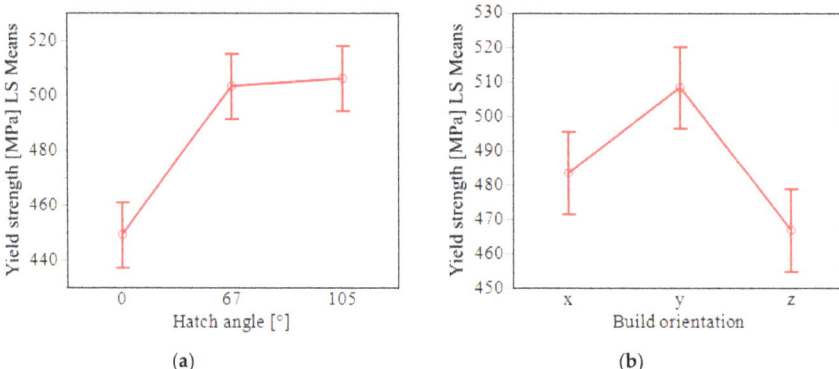

Figure 11. Least square mean plot (yield strength) for the two factors: (**a**) hatch angle, and (**b**) build orientation.

For the effects table for flow stress at 35% plastic strain (Table 10), the interaction term was found to be significant (i.e., the null hypothesis $H_{0I*II(35\% \text{ plastic})}$ was rejected). This means the plastic flow stress experienced during the compression of the samples was dependent on both the hatch angle and build orientation. Given the presence of interaction between these factors, a Tukey's test (Table 12) was conducted on the treatment combination means to determine which combination of factor levels produced the specimen with highest flow stress at 35% plastic strain. From the Tukey's test and the least squares means plot (Figure 12), it can be deduced that samples built with hatch angles of 67° and 105° produced parts which flowed at statistically the same stress at 35% plastic strain, while samples built with a 0° hatch angle flowed at a lower stress.

Table 12. LSMeans Differences Tukey HSD (flow stress at 35% plastic strain); $\alpha = 0.050$.

Level				Least Sq. Mean
105,z	A			1003.8160
105,y	A			994.5960
67,y	A			993.2092
105,x	A			992.9778
67,z	A			988.9414
67,x	A			987.4619
0,x		B		910.5729
0,y		B	C	876.1423
0,z			C	836.9911

Levels not connected by same letter are significantly different.

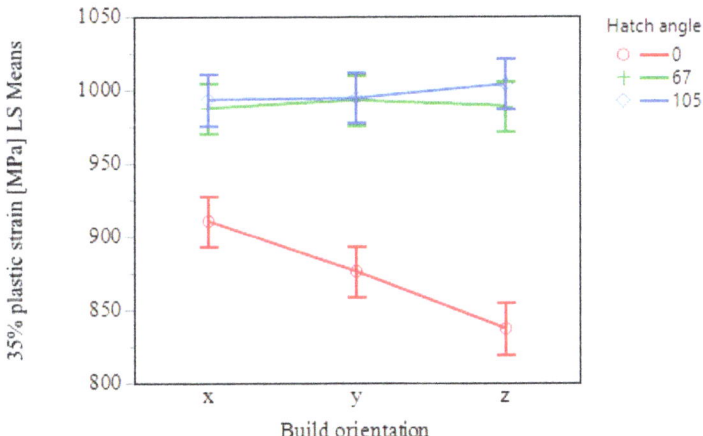

Figure 12. Least squares mean plot (flow stress at 35% plastic strain).

4. Conclusions

The objective of this study was to evaluate the best hatch angle and build orientation for manufacturing parts using SLM. Samples were manufactured using three hatch angles (0°, 67°, and 105°) in three build orientations (x-, y-, and z-direction), and tested in compression. The different compressive responses were evaluated using ANOVA and the Tukey's test. This study considered hatch angle and build orientation interaction, which made the comparison more accurate. The statistical analysis showed that changes in the hatch angle, and build direction caused changes in the measured yield strength and flow stress. Samples built in the y-direction exhibited the highest yield strength irrespective of the hatch angle; although samples manufactured using hatch angles 0° exhibited the lowest yield strength when compared to samples manufactured using the other two hatch angles. Samples manufactured with a 0° hatch angle flowed at the lowest stress at 35% plastic strain. Samples manufactured with hatch angles of 67° and 105° flowed at statistically the same flow stress at 35% plastic strain. However, it is important to note that under compression, samples built with a hatch angle of 67° deformed non-uniformly which is a source of concern. It was observed that samples built with a 0° hatch angle in all directions (x, y, and z) deformed non-uniformly under compressive loading while only samples built in the x- and z-direction with a 67° hatch angle deformed non-uniformly. Samples built with a 0° hatch angle deformed by ovalling when built in the x-direction and sheared when built in the z-direction, while samples built with a 67° hatch angle formed an ellipsoid when built in the x-direction and sheared when built in the z-direction. Only samples built with a 105° hatch angle deformed symmetrically whilst showing excellent compressive properties. Considering the results obtained from the compression, ANOVA, and the Tukey's test it can be concluded that 304L stainless steel parts manufactured in the y-direction using a 105° hatch angle showed the best overall compressive behavior.

Author Contributions: Conceptualization, K.C., J.N., H.M., M.W., and O.F.; methodology, O.F., M.S. and M.F.B.; formal analysis, O.F., and M.F.B.; investigation, O.F., K.C., and J.N.; resources, H.M. and M.W.; writing—original draft preparation, O.F., K.C., J.N. and M.F.B.; writing—review and editing, O.F., K.C., J.N., M.F.B., H.M., and M.W.; visualization, O.F., M.S. and M.F.B.; supervision, K.C. and J.N.; project administration, K.C.

Funding: This research received no external funding.

Acknowledgments: Support from Intelligent Systems Center (ISC) and Center for Aerospace Manufacturing Technologies (CAMT) at Missouri University of Science and Technology is gratefully acknowledged. The authors would also like to acknowledge Austin Sutton for his contributions, and Honeywell Federal Manufacturing and Technology for use of the Renishaw machine.

Conflicts of Interest: The authors declare no conflicts of interest.

References

1. Rajak, N.K.; Kaimkuriya, A. Design and Development of Honeycomb Structures for Additive Manufacturing. *Int. J. Trend Sci. Res. Develop. (IJTSRD)* **2018**, *2*, 1198–1203. [CrossRef]
2. Luisa, S.; Contuzzi, N.; Angelastro, A.; Domenico, A. Capabilities and Performances of the Selective Laser Melting Process. In *New Trends in Technologies: Devices, Computer, Communication and Industrial Systems*; IntechOpen: London, UK, 2010; pp. 233–252.
3. Guan, K.; Wang, Z.; Gao, M.; Li, X.; Zeng, X. Effects of Processing Parameters on Tensile Properties of Selective Laser Melted 304 Stainless Steel. *Mater. Design* **2013**, *50*, 581–586. [CrossRef]
4. Sames, W.J.; List, F.A.; Pannala, S.; Dehoff, R.R.; Babu, S.S. The Metallurgy and Processing Science of Metal Additive Manufacturing. *Int. Mater. Rev.* **2016**, *61*, 315–360. [CrossRef]
5. Brown, B.; Everhart, W.; Dinardo, J. Characterization of Bulk to Thin Wall Mechanical Response Transition in Powder bed AM. *Rapid Prototyp. J.* **2016**, *22*, 801–809. [CrossRef]
6. Popovich, A.A.; Sufiiarov, V.S.; Borisov, E.V.; Polozov, I.A.; Masaylo, D.V.; Grigoriev, A.V. Anisotropy of Mechanical Properties of Products Manufactured Using Selective Laser Melting of Powdered Materials. *Rus. J. Non-Ferrous Metals* **2017**, *58*, 389–395. [CrossRef]
7. Miranda, G.; Faria, S.; Bartolomeu, F.; Pinto, E.; Madeira, S.; Mateus, A.; Carreira, P.; Alves, N.; Silva, F.S.; Carvalho, O. Predictive Models for Physical and Mechanical Properties of 316L Stainless Steel Produced by Selective Laser Melting. *Mater. Sci. Eng.* **2016**, *657*, 43–56. [CrossRef]
8. Ortiz Rios, C.; Amine, T.L.; Newkirk, J.W. Tensile Behavior in Selective Laser Melting. *Int. J. Adv. Manuf. Technol.* **2018**, *96*, 1187–1194. [CrossRef]
9. Anam, M.A.; Dilip, J.J.S.; Pal, D.; Stucker, B. Effect of Scan Pattern on the Microstructural Evolution of Inconel 625 during Selective Laser Melting. In Proceedings of the Annual International Solid FreeForm Fabrication Symposium, Austin, TX, USA, 4–6 August 2014; pp. 363–376.
10. Wang, Z.; Palmer, T.A.; Beese, A.M. Effect of Processing Parameters on Microstructure and Tensile Properties of Austenitic Stainless Steel 304L Made by Directed Energy Deposition Additive Manufacturing. *Acta Mater.* **2016**, *110*, 226–235. [CrossRef]
11. Yadroitsev, I.; Bertrand, P.; Smurov, I. Parametric Analysis of the Selective Laser Melting Process. *Appl. Surface Sci.* **2007**, *253*, 8064–8069. [CrossRef]
12. Hanzl, P.; Zetek, M.; Bakša, T.; Kroupa, T. The Influence of Processing Parameters on the Mechanical Properties of SLM Parts. *Proc. Eng.* **2015**, *100*, 1405–1413. [CrossRef]
13. Kriewall, C.; Sutton, A.; Leu, M.; Newkirk, J.; Brown, B. Investigation of Heat-Affected 304L SS Powder and its Effect on Built Parts in Selective Laser Melting. In Proceedings of the Solid Freeform Fabrication Conference—An Additive Manufacturing Conference, Austin, TX, USA, 8–10 August 2016; pp. 363–376.
14. Karnati, S.; Axelsen, I.; Liou, F.F.; Newkirk, J.W. Investigation of Tensile Properties of Bulk and SLM Fabricated 304L Stainless Steel Using Various Gage Length Specimens. In Proceedings of the Solid Freeform Fabrication Conference—An Additive Manufacturing Conference, Austin, TX, USA, 8–10 August 2016; 592–604; pp. 592–604.
15. Song, B.; Nishida, E.; Sanborn, B.; Maguire, M.; Adams, D.; Carroll, J.; Wise, J.; Reedlunn, B.; Bishop, J.; Palmer, T. Compressive and Tensile Stress–Strain Responses of Additively Manufactured (AM) 304L Stainless Steel at High Strain Rates. *J. Dyn. Behav. Mater.* **2017**, *3*, 412–425. [CrossRef]
16. Hitzler, L.; Hirsch, J.; Heine, B.; Merkel, M.; Hall, W.; Öchsner, A. On the Anisotropic Mechanical Properties of Selective Laser-Melted Stainless Steel. *Materials* **2017**, *10*, 1136. [CrossRef] [PubMed]
17. Read, N.; Wang, W.; Essa, K.; Attallah, M.M. Selective laser melting of AlSi10Mg: Process optimisation and mechanical properties development. *Mater. Design* **2015**, *65*, 417–424. [CrossRef]
18. Calignano, F.; Manfredi, E.P.D.A.; Iuliano, L.; Fino, P. Influence of process parameters on surface roughness of aluminum parts produced by DMLS. *Int. J. Adv. Manuf. Technol.* **2013**, *67*, 2743–2751. [CrossRef]
19. ASTM Standard 52921. Standard Terminology for Additive Manufacturing—Coordinate Systems and Test. In *Annual ASTM Standard 2015*; ASTM International: West Conshohocken, PA, USA, 2015; Volume 2013, pp. 1–13.
20. Alsalla, H.; Hao, L.; Smith, C.W. Effect of build orientation on the surface quality, microstructure and mechanical properties of selective laser melting 316L stainless steel. *Rapid Prototyp. J.* **2017**, *24*, 9–17. [CrossRef]

21. Riemer, B.W. Benchmarking dynamic strain predications of pulsed mercury spallation target vessels. *J. Nucl. Mater.* **2005**, *343*, 81–91. [CrossRef]
22. ASTM Standard E9-09. Standard Test Methods of Compression Testing of Metallic Materials at Room Temperature. In *Annual ASTM Standard 2012*; ASTM International: West Conshohocken, PA, USA, 2012; Volume 3.01, pp. 92–100.
23. ASTM Standard E3-11. Standard Guide for Preparation of Metallographic Specimens 1. In *Annual ASTM Standard 2001*; ASTM International: West Conshohocken, PA, USA, 2001; Volume 03, pp. 1–17.
24. Montgomery, D.C. *Design and Analysis of Experiments*, 5th ed.; John Wiley & Sons Inc.: Hoboken, NJ, USA, 2005.
25. Amine, T.; Kriewall, C.S.; Newkirk, J.W. Long-Term Effects of Temperature Exposure on SLM 304L Stainless Steel. *J. Miner. Metals Mater. Soc.* **2018**, *70*, 384–389. [CrossRef]
26. Li, L.; Lough, C.; Replogle, A.; Bristow, D.; Landers, R.; Kinzel, E. Thermal Modeling of 304L Stainless Steel Selective Laser Melting. In Proceedings of the Solid Freeform Fabrication Conference—An Additive Manufacturing Conference, Austin, TX, USA, 7–9 August 2017; pp. 1068–1081.
27. Wang, Y.M.; Voisin, T.; McKeown, J.T.; Ye, J.; Calta, N.P.; Li, Z.; Zeng, Z.; Zhang, Y.; Chen, W.; Roehling, T.T. Additively manufactured hierarchical stainless steels with high strength and ductility. *Nat. Mater.* **2017**, *17*, 63–71. [CrossRef] [PubMed]
28. Qiu, C.; Al Kindi, M.; Aladawi, A.S.; Al Hatmi, I. A comprehensive study on microstructure and tensile behaviour of a selectively laser melted stainless steel. *Sci. Rep.* **2018**, *8*, 1–16. [CrossRef] [PubMed]
29. Meier, H.; Haberland, C. Experimental Studies on Selective Laser Melting of Metallic Parts. *Materialwiss. Werkstofftech.* **2008**, *39*, 665–670. [CrossRef]

© 2019 by the authors. Licensee MDPI, Basel, Switzerland. This article is an open access article distributed under the terms and conditions of the Creative Commons Attribution (CC BY) license (http://creativecommons.org/licenses/by/4.0/).

Article

Phase Change with Density Variation and Cylindrical Symmetry: Application to Selective Laser Melting

Marios M. Fyrillas [1,†], Yiannos Ioannou [2], Loucas Papadakis [1,*], Claus Rebholz [3,4], Allan Matthews [4] and Charalabos C. Doumanidis [5,6]

1. Department of Mechanical Engineering, Frederick University, Nicosia 1303, Cyprus
2. Department of Physical Metallurgy and Materials Testing, Montanuniversität Leoben, Leoben 8700, Austria
3. Department of Mechanical and Manufacturing Engineering, University of Cyprus, Nicosia 1678, Cyprus
4. School of Materials, University of Manchester, ICAM-Pariser Building, Manchester M13BB, UK
5. Office of the Provost, Nazarbayev University, Astana 010000, Kazakhstan
6. College of Engineering and Computer Science, Vin University, Hanoi, Vietnam
* Correspondence: l.papadakis@frederick.ac.cy
† Author deceased.

Received: 17 June 2019; Accepted: 22 July 2019; Published: 25 July 2019

Abstract: In this paper we introduce an analytical approach for predicting the melting radius during powder melting in selective laser melting (SLM) with minimum computation duration. The purpose of this work is to evaluate the suggested analytical expression in determining the melt pool geometry for SLM processes, by considering heat transfer and phase change effects with density variation and cylindrical symmetry. This allows for rendering first findings of the melt pool numerical prediction during SLM using a quasi-real-time calculation, which will contribute significantly in the process design and control, especially when applying novel powders. We consider the heat transfer problem associated with a heat source of power $\dot{Q}'$ (W/m) per unit length, activated along the span of a semi-infinite fusible material. As soon as the line heat source is activated, melting commences along the line of the heat source and propagates cylindrically outwards. The temperature field is also cylindrically symmetric. At small times (i.e., neglecting gravity and Marangoni effects), when the density of the solid material is less than that of the molten material (i.e., in the case of metallic powders), an annulus is created of which the outer interface separates the molten material from the solid. In this work we include the effect of convection on the melting process, which is shown to be relatively important. We also justify that the assumption of constant but different properties between the two material phases (liquid and solid) does not introduce significant errors in the calculations. A more important result; however, is that, if we assume constant energy input per unit length, there is an optimum power of the heat source that would result to a maximum amount of molten material when the heat source is deactivated. The model described above can be suitably applied in the case of selective laser melting (SLM) when one considers the heat energy transferred to the metallic powder bed during scanning. Using a characteristic time and length for the process, we can model the energy transfer by the laser as a heat source per unit length. The model was applied in a set of five experimental data, and it was demonstrated that it has the potential to quantitatively describe the SLM process.

Keywords: selective laser melting (SLM); analytical melt pool calculation; phase change; cylindrical symmetry; line heat source

1. Introduction

Consider a semi-infinite solid slab, initially at temperature T_∞. At time $t = 0$, a continuous line heat source $\dot{Q}'\,(W/m)$ is activated along the line $r = 0$. The temperature distribution in the semi-infinite medium is found to be [1] (pg. 261, Equation (5))

$$T = T_\infty + \frac{\dot{Q}'}{2\pi k} E_1\left[\frac{r^2}{4\alpha t}\right] \quad (1)$$

where k is the thermal conductivity of the solid, and $E_1[x]$ is the exponential integral defined as $E_1[x] = \int_x^\infty e^{-t}/t\, dt$. Similar to 1D Cartesian coordinates, there is no steady-state solution in 1D cylindrical coordinates. The temperature field is cylindrically symmetric, ranging from infinite at $r = 0$ to T_∞ at the far field. In reality though, if T_∞ is less than the melting temperature T_{melt} of the material, the high temperature developed around $r = 0$ would lead to melting of the material and a cylindrical interface would emerge separating the solid phase (metal powder) from the liquid phase (molten material). Hence, one has to take into consideration the different properties between the liquid and the solid phases and, in addition, the latent heat of melting L per unit mass at the interface. For a pure substance the interface is sharp at the melting temperature T_{melt} of the material, and it moves cylindrically outwards separating the two phases, solid and liquid. Similar arguments apply for the solidification process.

For the case of melting or solidification of a material due to a line source/sink, with constant but different properties between the liquid and solid phases but equal densities, an analytical solution has been obtained by Patterson [2], who combined two expressions similar to Equation (1). The boundary condition for the energy balance leads to an algebraic equation (characteristic equation) where the unknown, which can be considered to be an eigenvalue, is proportional to the position of the liquid-solid interface (i.e., the speed of melting of the material). The characteristic equation is monotonic with respect to the eigenvalue so there is only one single solution. The analysis is valid when the two phases have the same density. A review is given by Hu and Argyropoulos [3] and Alexiades and Solomon [4], where they present the major methods of mathematical modeling of solidification and melting. For the case where the two phases have different densities, the energy balance equation is different [4,5], and furthermore a convection term must be included in the heat equation of the liquid phase. The convection term can be obtained from the radially-symmetric continuity equation in cylindrical coordinates with constant density [4]. A similar approach is used for the analysis of melting of nanoparticles [6] and bubble growth and oscillations [7–11].

The major difference between bubble growth/oscillations and melting/solidification is that, for the former case, the bubble interface is set in motion by the pressure field, hence one has to solve the momentum equation. On the contrary, for the latter case the interface between the liquid and the solid phase is controlled by the conduction process; hence, the heat equation is decoupled from the momentum equation.

In this work, similar to Font et al. [6], we first solve the continuity equation in the liquid phase to find an expression for the velocity field using mass conservation. Subsequently we solve the heat equation in the two phases using a similarity transformation of the form $r/\sqrt{t}$, and through the energy boundary condition on the interface we obtain an algebraic (characteristic) equation for the eigenvalue λ, which is proportional to the location of the interface. Unlike Font et al. [6], we have neglected the kinetic energy term in the energy equation because it is small compared to other terms (see Section 2.2). Furthermore, if needed, an expression for the pressure field in the liquid phase can be obtained by substituting the expression of the velocity field in the radial momentum equation.

In this work, we apply the melting process described above, as a simplified model to describe the dynamics of selective laser melting (SLM) processes. Of course, for the prediction of the melt pool when a particular SLM or selective laser sintering (SLS) process is concerned, different mathematical approaches have been introduced and studied in the literature. Cheng and Chou [12] described an unsteady temperature simulation based on the finite element method (FEM) of the alloy IN718 [13], concentrating on the effect of varying scan length on the melt pool size. Polivnikova [14] studied the melt pool dynamics by means of a finite-element simulation for an SLS/SLM process. The numerical model considered the interaction between laser beam and powder material and phase transformations, while sub-models were developed to describe the capillary phenomena in the powder bed during SLS/SLM processing. Li et al. [15] investigated the heat transfer and phase transition during an SLM process with a moving volumetric heat source using the finite difference method. They proposed a model incorporating a phase function to differentiate the powder phase, melting liquid phase, dense solid phase and vaporized gas phase that also includes the volume shrinkage induced by the density change during the melting process. Letenneur et al. proposed a three dimensional analytical model which enables the calculation of the temperature distribution in powder for a Gaussian laser heat source [16]. In [17], Li et al. enhanced their proposed approach by including the residual stress field analysis. A similar numerical approach was presented by Tan et al. [18] based on a model addressing thermal, metallurgical, and mechanical effects for selective laser melting of titanium alloy. The aforementioned numerical models, although they consider all physical phenomena and elucidate the physical processes involved in the melt pool formation, are computationally intensive and cannot be used for real-time process control and optimization.

Thus, for the design of a thermal control system, the development of an efficient model is inevitable, and an analytical or semi-analytical model is necessary. On one hand, it should contain all the necessary physics, and on the other hand it should be characterized by the necessary computational efficiency, so that it can be used as an in-process reference for a control algorithm. Examples are available in the literature for pure heat transfer operations, such as scanned thermal processing [19,20] and for serial thermal processing methods such as arc welding [21] and SLM and laser cutting [22]. The model can be validated through an experimental apparatus via infrared camera and laser profilometry. This kind of sensor was used for output feedback in a closed loop geometry control system in a gas metal arc welding (GMAW) process [23]. Besides control, a simplified model can also be used to compute the structural shape and residual stresses [24]. In what follows, we develop a simple model based on the melting achieved due to a line heat source.

2. Mathematical Modeling

Our model is motivated by the melting achieved by a rapidly scanning laser beam moving along the span of a powder-bed [25]. Figures 1 and 2 provide a detailed description of the model.

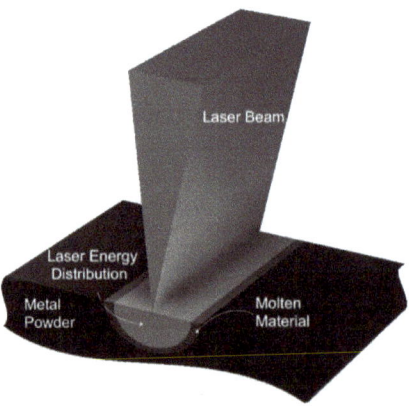

Figure 1. Heat source (light gray color) of strength $\dot{Q}'$ (W/m) distributed along the inner surface of the molten material as power per unit length.

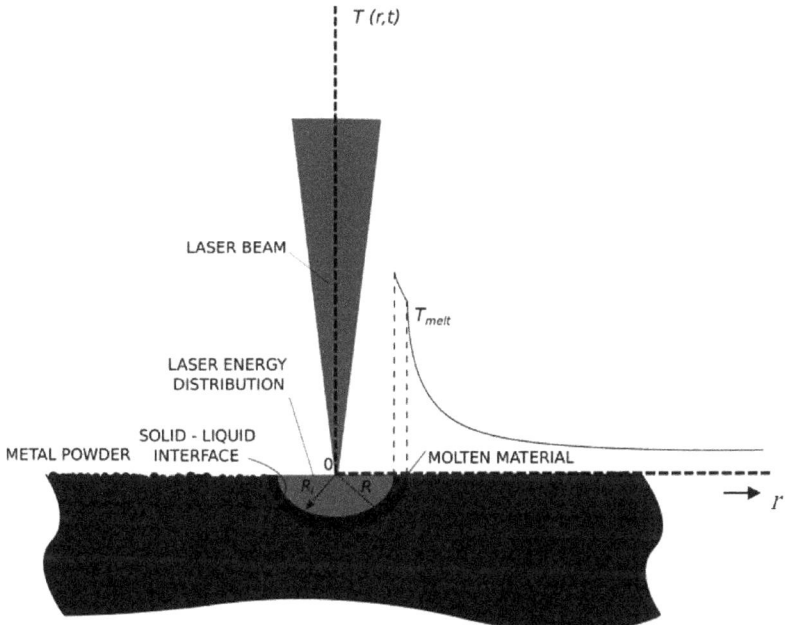

Figure 2. A semi-infinite slab of fusible material melts cylindrically due to a heat source $\dot{Q}'(W/m)$. The temperature field is cylindrically symmetric. The uniformly cylindrical power distribution from the laser source (gray color) is assumed over the inner surface of the molten material with radius R_i.

Initially, the semi-infinite bed of the fusible material (metal powder, powder-bed) is at the temperature T_∞, lower than the melting temperature T_{melt} of the powder. A line heat source of strength $\dot{Q}'(W/m)$ is distributed initially along $r = 0$ and is activated for a brief period of time τ. The release of energy causes the temperature to increase higher than T_{melt}, hence the melting commences at the origin. Because the density of the liquid phase is higher than the density of the solid phase (powder), an inner interface is created with radius R_i (liquid-air interface, Figure 2). Hence, an annulus is created with an inner surface of radius $R_i[t]$, and an outer surface of radius $R[t]$. The latter separates the liquid (melted material) from the solid (powder) (i.e., $T[r = R[t]] = T_{melt}$; Figure 2); assuming a pure substance, and by ignoring the kinetics of phase change in melting, the interface ($r = R[t]$) is sharp. Both the inner $r = R_i[t]$ and the outer $r = R[t]$ interfaces move in the positive r-direction with cylindrical symmetry, as shown in Figures 1 and 2. During the time period $0 < t \leq \tau$, we assume that the laser beam is distributed evenly over the interface $R_i[t]$ due to irradiation and reflections of the laser beam, whereas at $t = 0$ the laser beam is a line heat source distributed along $r = 0$. For small times we can neglect gravity and Marangoni effects, hence the process is cylindrically symmetric, and, in addition, a radially-symmetric convection current $u_r[r,t]$ is developed in the annulus.

In order to apply this model to SLM of metallic powders a number of assumptions/parameters were adopted as described in detail in the following sections. The metal powder, which rests on a metal substrate, is molten by the scanning laser beam and solidifies into a metal deposit fused to the substrate. The density and thermal properties of the deposit are significantly different from the metal powder (see beginning of Section 3). We assume that during the initial stages of melting, the gravity and Marangoni effects can be neglected, hence the process is cylindrically symmetric. Furthermore, because the conductivity of the metal powder is on the order of 100 times smaller than that of the metal substrate, we assume that the heat transfer process is controlled by the powder bed, and the heat transfer process is prevalent in the molten pool [1]. Hence, during the initial melting of the powder, we neglect the presence of the metal substrate and assume a semi-infinite layer of metal powder.

The analysis that follows proceeds along the same lines as the analysis by Font et al. [6] for melting of a nanoparticle, and the analysis by Scriven [9] for bubble growth and oscillations [7–11], where an explicit expression for the convection term is obtained from the continuity equation and mass conservation.

2.1. Flow Field

Assuming constant density in the liquid phase (i.e., zero thermal expansion coefficient), the equation of continuity in cylindrical coordinates [26] takes the form $\frac{1}{r}\frac{\partial}{\partial r}(\rho_\ell r u_r) = 0$. Integrating with respect to r, and enforcing the conditions at $r = R[t]$ or $r = R_i[t]$ we obtain:

$$u_r[r,t] = \frac{R[t]\, v_\ell}{r} = \frac{R_i[t]\, \dot{R}_i[t]}{r} \tag{2}$$

where $u_r[r,t]$ is the velocity of the fluid at radius r and time t and $v_\ell \equiv u_r[r = R[t], t]$ is the velocity of the fluid perpendicular to the interface $r = R[t]$. Similar expressions are obtained in the case of a moving spherical interface [6–11]. We should point out that, while the velocity of the fluid (molten material) is $\dot{R}_i[t]$ at $r = R_i$, its velocity is not $\dot{R}[t]$ at $r = R[t]$. In order to find a relation for v_ℓ we use mass conservation in a frame moving with the interface to obtain

$$-\rho_s \dot{R}[t] = \left(v_\ell - \dot{R}[t]\right)\rho_\ell,$$

where ρ_s is the density of the solid and ρ_ℓ is the density of the liquid at the melting temperature T_{melt}, which can be is resolved to

$$v_\ell = \frac{(\rho_\ell - \rho_s)}{\rho_\ell}\dot{R}[t]. \tag{3}$$

This is the equation obtained by Özişik [5] (pg. 403, Equation (10-9b)). To relate $R_i[t]$ to $R[t]$, we use mass conservation [6] on a unit span of the annulus of the molten pool to obtain

$$\frac{1}{2}\frac{d}{dt}\left[\rho_\ell \pi \left(R[t]^2 - R_i[t]^2\right)\right] = \pi \rho_s R[t]\dot{R}[t].$$

Integrating above expression we obtain

$$\rho_\ell \left(R[t]^2 - R_i[t]^2\right) = \rho_s R[t]^2$$
$$\Rightarrow R[t]\sqrt{\frac{\rho_\ell - \rho_s}{\rho_\ell}} = R_i[t] \tag{4}$$

Substituting Equation (3) or Equation (4) in Equation (2), we obtain an equation for the velocity field in terms of the location $R[t]$ and velocity $\dot{R}[t]$ of the interface:

$$u_r[r,t] = \frac{(\rho_\ell - \rho_s)}{\rho_\ell}\frac{R[t]\,\dot{R}[t]}{r} \tag{5}$$

2.2. Governing Equations

Assuming that the properties of the powder and the molten material are constant but different, the mathematical formulation of this problem is given by

$$\begin{aligned}\rho_\ell\, c_{p\ell}\left(\frac{\partial T_\ell}{\partial t} + u_r[r,t]\frac{\partial T_\ell}{\partial r}\right) &= \frac{1}{r}\frac{\partial}{\partial r}\left(k_\ell\, r\, \frac{\partial T_\ell}{\partial r}\right) \text{ in the region } R_i[t] < r < R[t],\ t > 0 \\ \rho_s\, c_{ps}\frac{\partial T_s}{\partial t} &= \frac{1}{r}\frac{\partial}{\partial r}\left(k_s\, r\, \frac{\partial T_s}{\partial r}\right) \text{ in the region } R[t] < r < \infty,\ t > 0\end{aligned} \tag{6}$$

where, as mentioned earlier (Equation (5)),

$$u_r[r,t] = \frac{(\rho_\ell - \rho_s)}{\rho_\ell} \frac{R[t] \, \dot{R}[t]}{r}$$

with the initial condition

$$T = T_\infty \text{ at } t = 0$$

and the boundary conditions

$$\begin{aligned}
-\pi \, r \, k_\ell \frac{\partial T_\ell}{\partial r} &= \dot{Q}' \text{ at } r = R_i[t], \; t > 0, \\
T_s &= T_\ell = T_{\text{melt}} \text{ at } r = R[t], \; t > 0, \\
k_s \frac{\partial T_s}{\partial r} - k_\ell \frac{\partial T_\ell}{\partial r} &= L \, \rho_s \frac{dR}{dt} \text{ at } r = R[t], \; t > 0 \\
T_s &\to T_\infty \text{ as } r \to \infty, \; t > 0.
\end{aligned} \quad (7)$$

In the above equations, ρ represents the density, c_p the specific heat, and k the conductivity, while the subscripts ℓ and s represent the liquid and the solid phases, respectively. As shown in Özişik [5] (pg. 403, Equation (10-10a)), the third boundary condition is developed by performing an energy balance across the melting interface at $r = R[t]$. In a coordinate system moving with the interface the energy balance takes the form:

$$\dot{R}[t] \, H_s \, \rho_s + k_s \frac{\partial T_s}{\partial r} = \left((\dot{R}[t] - v_\ell) H_\ell \rho_\ell + k_\ell \frac{\partial T_\ell}{\partial r} \right)$$

If we substitute Equation (3) for v_ℓ we obtain

$$k_s \frac{\partial T_s}{\partial r} - k_\ell \frac{\partial T_\ell}{\partial r} = \dot{R}[t] \left(H_\ell \, \rho_\ell - \frac{(\rho_\ell - \rho_s)}{\rho_\ell} H_\ell \, \rho_\ell - H_s \, \rho_s \right)$$

$$\Rightarrow k_s \frac{\partial T_s}{\partial r} - k_\ell \frac{\partial T_\ell}{\partial r} = \dot{R}[t] \, (H_\ell \, \rho_\ell - (\rho_\ell - \rho_s) H_\ell - H_s \, \rho_s)$$

which is finally simplified to

$$\Rightarrow k_s \frac{\partial T_s}{\partial r} - k_\ell \frac{\partial T_\ell}{\partial r} = \dot{R}[t] \, \rho_s \, (H_\ell - H_s) = \dot{R}[t] \, \rho_s \, L,$$

where L is the latent heat, $L = H_\ell - H_s$.

A common mistake that appears in phase change problems with density variations is the exclusion of the kinetic energy term [4,6]. This term is the consequence of the density change which forces the fluid to move, and results in a kinetic energy deficit or surplus. This term is equal to $\pm \frac{\rho_s}{2} \left(1 - \frac{\rho_s}{\rho_\ell}\right)^2 \left(\dot{R}[t]\right)^3$ and it is usually excluded if it is small compared to the term related to the latent heat $\left(\dot{R}[t] \, \rho_s \, L\right)$. It is important only at very small times and when the value of the latent heat L is small [4]; an exception is the melting of nanoparticles [6]. In our simulation the latent heat is of the order $L \sim 10^5$ (J/kg) and the smaller value of the activation time $\tau \sim 0.0001$ (s). As we will show later $\dot{R}[t] = \frac{\lambda \sqrt{\alpha_\ell}}{\sqrt{t}}$, hence the kinetic energy term is of the order $\sim 10^{-2}$ (kg/s^3) and it does not introduce any significant error. The important advantage of neglecting this term is that it allows for a similarity solution for the system of equations [2,4,5] in the form $z = r/R[t]$. The partial derivatives transform as follows:

$$\frac{\partial}{\partial r} = \frac{1}{R[t]} \frac{\partial}{\partial z}, \; \frac{\partial}{\partial t} = \frac{\partial}{\partial t} - \frac{r \, \dot{R}[t]}{R[t]^2} \frac{\partial}{\partial z}.$$

Substituting in the partial differential Equations (6) and the boundary conditions (7), we obtain the following system of ordinary differential equations, which we assume that they are independent of time (t):

$$\rho_\ell c_{p\ell} \left(-\frac{r \dot{R}[t]}{R[t]^2} \frac{dT_\ell}{dz} + \frac{u_r[r,t]}{R[t]} \frac{dT_\ell}{dz} \right) = \frac{1}{r R[t]} \frac{d}{dz} \left(k_\ell \frac{r}{R[t]} \frac{dT_\ell}{dz} \right) \text{ for } \frac{R_i[t]}{R[t]} < z < 1$$

$$\rho_s c_{ps} \left(-\frac{r \dot{R}[t]}{R[t]^2} \frac{dT_s}{dz} \right) = \frac{1}{r R[t]} \frac{d}{dz} \left(k_s \frac{r}{R[t]} \frac{dT_s}{dz} \right) \text{ for } 1 < z < \infty$$

with boundary conditions

$$-\pi \frac{r}{R[t]} k_\ell \frac{dT_\ell}{dz} = \dot{Q}' \text{ at } z = \frac{R_i[t]}{R[t]},$$

$$T_s = T_\ell = T_{\text{melt}} \text{ at } z = 1$$

$$\frac{k_s}{R[t]} \frac{dT_s}{dz} - \frac{k_\ell}{R[t]} \frac{dT_\ell}{dz} = L \rho_s \frac{dR}{dt} \text{ at } r = 1,$$

$$T_s \to T_\infty \text{ at } z \to \infty.$$

Note that the last equation also describes the initial condition. Multiplying by $R[t]^2$ and substituting the expression for $u_r[r,t] = \frac{(\rho_\ell - \rho_s)}{\rho_\ell} \frac{R[t] \dot{R}[t]}{r}$ (Equation (4)) and $R_i[t] = R[t] \sqrt{\frac{\rho_\ell - \rho_s}{\rho_\ell}}$ (Equation (4)), the above system simplifies to

$$\rho_\ell c_{p\ell} \left(-\frac{r R[t] \dot{R}[t]}{R[t]} \frac{dT_\ell}{dz} + \frac{R[t]^2 (\rho_\ell - \rho_s)}{\rho_\ell} \frac{\dot{R}[t]}{r} \frac{dT_\ell}{dz} \right) = \frac{R[t]}{r} \frac{d}{dz} \left(k_\ell \frac{r}{R[t]} \frac{dT_\ell}{dz} \right) \text{ for } \sqrt{\frac{\rho_\ell - \rho_s}{\rho_\ell}} < z < 1$$

$$\rho_s c_{ps} \left(-\frac{r R[t] \dot{R}[t]}{R[t]} \frac{dT_s}{dz} \right) = \frac{R[t]}{r} \frac{d}{dz} \left(k_s \frac{r}{R[t]} \frac{dT_s}{dz} \right) \text{ for } 1 < z < \infty$$

with boundary conditions

$$-\pi z k_\ell \frac{dT_\ell}{dz} = \dot{Q}' \text{ at } z = \sqrt{\frac{\rho_\ell - \rho_s}{\rho_\ell}},$$

$$T_s = T_\ell = T_{\text{melt}} \text{ at } z = 1,$$

$$k_s \frac{dT_s}{dz} - k_\ell \frac{dT_\ell}{dz} = L \rho_s R[t] \frac{dR}{dt} \text{ at } z = 1,$$

$$T_s \to T_\infty \text{ as } z \to \infty.$$

Above equations are independent of time only if $R[t]\dot{R}[t] = c$ (i.e., $R[t] = \sqrt{2ct}$, where c is a constant to be determined). We finally obtain:

$$\rho_\ell c_{p\ell} \left(-zc \frac{dT_\ell}{dz} + \frac{(\rho_\ell - \rho_s)}{\rho_\ell} \frac{c}{z} \frac{dT_\ell}{dz} \right) = \frac{1}{z} \frac{d}{dz} \left(k_\ell z \frac{dT_\ell}{dz} \right) \text{ for } \sqrt{\frac{\rho_\ell - \rho_s}{\rho_\ell}} < z < 1$$

$$\rho_s c_{ps} \left(-zc \frac{dT_s}{dz} \right) = \frac{1}{z} \frac{d}{dz} \left(k_s[T] z \frac{dT_s}{dz} \right) \text{ for } 1 < z < \infty$$

with boundary conditions

$$-\pi z k_\ell[T] \frac{dT_\ell}{dz} = \dot{Q}' \text{ at } z = \sqrt{\frac{\rho_\ell - \rho_s}{\rho_\ell}},$$

$$T_s = T_\ell = T_{\text{melt}} \text{ at } z = 1$$

$$k_s[T] \frac{dT_s}{dz} - k_\ell[T] \frac{dT_\ell}{dz} = L \rho_s[T] c \text{ at } z = 1,$$

$$T_s \to T_\infty \text{ as } z \to \infty,$$

where the square brackets $[T]$ indicate a temperature dependency of the thermal conductivity k and the density ρ. The above system can be brought in the following form:

$$\frac{d^2 T_\ell}{dz^2} + \left(\frac{1}{z} + \frac{zc}{\alpha_\ell} - \frac{\varepsilon}{\alpha_\ell}\frac{c}{z}\right)\frac{dT_\ell}{dz} = 0 \text{ for } \sqrt{\varepsilon} < z < 1$$

$$\frac{d^2 T_s}{dz^2} + \left(\frac{1}{z} + \frac{zc}{\alpha_s}\right)\frac{dT_s}{dz} = 0 \text{ for } 1 < z < \infty \tag{8}$$

with boundary conditions

$$-\pi z k_\ell \frac{dT_\ell}{dz} = \dot{Q}' \text{ at } z = \sqrt{\varepsilon},$$

$$T_s = T_\ell = T_{melt} \text{ at } z = 1,$$

$$k_s \frac{dT_s}{dz} - k_\ell \frac{dT_\ell}{dz} = L\rho_s c \text{ at } z = 1, \tag{9}$$

$$T_s \to T_\infty \text{ as } z \to \infty.$$

where $\alpha = k/(\rho c_p)$ is the thermal diffusivity and $\varepsilon = (\rho_\ell - \rho_s)/\rho_\ell$. Note that if we set $\varepsilon = 0$, the effect of convection is "switched off"; however, the density difference is still included in the energy balance equation (Equation (9), third equation; i.e., we obtain a result similar to Özişik [5] (pg. 415).

2.3. Characteristic Equation

The system of ordinary differential equations (ODEs) (8) can be integrated once to obtain

$$\frac{dT_\ell}{dz} = A_1 e^{-cz^2/(2\alpha_\ell)} z^{(c\varepsilon/\alpha_\ell - 1)},$$

$$\frac{dT_s}{dz} = A_2 e^{-cz^2/(2\alpha_s)} z^{-1}. \tag{10}$$

Using the first boundary condition from Equations (9) we obtain that

$$A_1 = -\frac{\dot{Q}' e^{\frac{c\varepsilon}{2\alpha_\ell}} \varepsilon^{-\frac{c\varepsilon}{2\alpha_\ell}}}{\pi k_\ell} = -\frac{\dot{Q}'}{\pi k_\ell}\left(\frac{e}{\varepsilon}\right)^{\frac{c\varepsilon}{2\alpha_\ell}}. \tag{11}$$

An expression for T_ℓ can be obtained in the form of the incomplete gamma function and the second boundary condition (Equation (9)); however, it is not required in order to obtain an expression for the eigenvalue c. An expression for T_s can be obtained in the form of the exponential integral $E_1[x] = \int_x^\infty e^{-t}/t \, dt$ using the substitution $\zeta = cz^2/(2\alpha_s)$:

$$\frac{dT_s}{d\zeta} = \frac{A_2}{2} e^{-\zeta}/\zeta.$$

Employing the second and fourth boundary conditions we can obtain an expression for A_2:

$$A_2 = \frac{2(T_\infty - T_{melt})}{E_1[c/(2\alpha_s)]} \tag{12}$$

Substituting (10)–(12) in the third boundary condition of Equation (9), we obtain an algebraic equation (i.e., the characteristic equation) for c:

$$\frac{\dot{Q}'}{\pi}\left(\frac{e}{\varepsilon}\right)^{c\varepsilon/(2\alpha_\ell)} e^{-c/(2\alpha_\ell)} + k_s \frac{2(T_\infty - T_{melt})}{E_1[c/(2\alpha_s)]} e^{-c/(2\alpha_s)} = L\rho_s c$$

If we replace c with $c = 2\,\alpha_\ell\,\lambda^2$, we obtain a result similar to Carslaw and Jaeger [26] (pg. 296):

$$R[t] = 2\,\lambda\,\sqrt{\alpha_\ell\,t},$$

$$F[\lambda] = \frac{Q'}{\pi}e^{-\lambda^2}\left(\frac{e}{\varepsilon}\right)^{(\lambda^2\varepsilon)} + \frac{2\,k_s\,(T_\infty - T_{melt})e^{(-\lambda^2\alpha_\ell/\alpha_s)}}{E_1[\lambda^2\,\alpha_\ell/\alpha_s]} - 2\,L\,\rho_s\,\lambda^2\,\alpha_\ell = 0. \quad (13)$$

The factor 1/2, instead of 1/4, is due to the fact that we have considered a semi-infinite domain instead of an infinite domain. We can reproduce the result by Carslaw and Jaeger [1] if we neglect the convection term (i.e., if we set $\varepsilon = 0$, and replace $E_1[x]$ with $E_1[x] = -Ei[-x]$). If we know the properties of the material and the power Q' per unit length of the heat source, the above equation can be solved to find λ; hence the location of the liquid-solid interface $R[t]$ can be obtained for the time period τ that the heat source is activated.

3. Numerical Results of the Characteristic Equation

3.1. Application for Material Properties of IN718 Powder

The final result of the previous section was the transcendental equation (characteristic Equation (13)) for the constant λ, through which we obtained the velocity of the interface (i.e., the speed of melting of the material). Equation (13) is monotonic so there is only one single value for the eigenvalue λ that satisfies the equation [2,5]. Furthermore, the effect of convection was to enhance the effect of the power Q', because the term $(e/\varepsilon)^{(\lambda^2\varepsilon)}$ is always greater than one. Hence, we expect that the effect of convection would lead to larger values of λ and, consequently, of the radius $R[t]$ of the melted material. The thermophysical properties of solid and liquid IN718 alloy were taken from the literature [12,13,27]. For our calculations we used the following values for the properties of the material which resembled the average IN718 powder properties for temperature close to melting point (i.e., in the range 800–1000 °C:

$$\rho_\ell = 7756\ \text{kg/m}^3,\ c_{p\ell} = 643\ \text{J/(kg·K)},\ k_\ell = 26.63\ \text{W/(m·K)},$$

$$\rho_s = 3926\ \text{kg/m}^3,\ c_{ps} = 351\ \text{J/(kg·K)},\ k_s = 0.37\ \text{W/(m·K)},$$

$$T_{melt} = 1300\ °\text{C},\ T_\infty = 20\ °\text{C},\ L = (643 - 351) \times (1300 + 273.15)\ \text{J/kg} = 459360\ \text{J/kg}\,,\ \varepsilon = 0.4934.$$

The melting temperature $T_{melt} = 1300\ °\text{C}$ used to model the melt pool was averaged from the liquid-solid phase distribution ranging from 1260 to 1336 °C [27]. An experimental measurement of enthalpy as a function of temperature indicating the temperature range for the liquid-solid phase change is shown in Figure 3 [28].

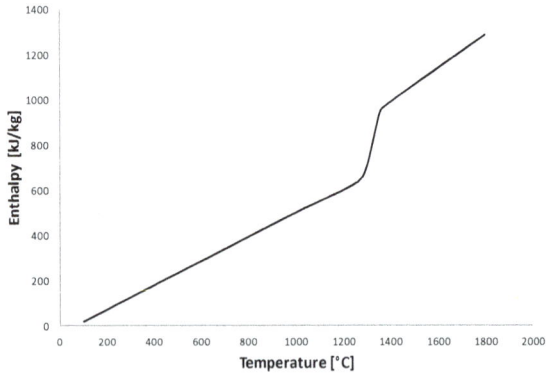

Figure 3. Liquid-solid phase distribution for IN718 powder shown on the basis of the enthalpy change as a function of temperature based on data from [27,28].

As an example, we set the power $\dot{Q}' = 10000$ W/m. In Figure 4, we show a plot of the function $F[\lambda]$ vs λ. As mentioned earlier, the function $F[\lambda]$ is monotonic and it is straightforward to obtain numerically the root $\lambda = 0.255$, hence the radius of the interface is $R[t] = 2\lambda \sqrt{a_\ell t} = 0.0012 \sqrt{t}$ m. This expression is valid while the heat source is activated (i.e., $0 \le t \le \tau$).

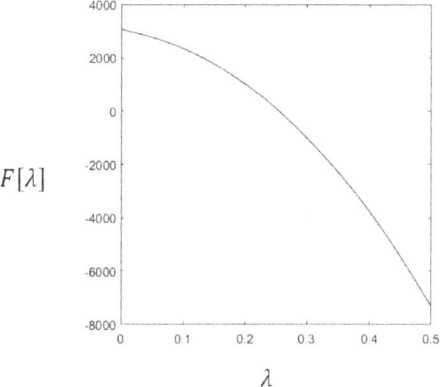

Figure 4. The characteristic Equation (13) as a function of the eigenvalue λ. The root is at the point where $F[\lambda] = 0$ (i.e., $\lambda = 0.255$).

To evaluate the effect of convection we set the parameter of the density ratio ε equal to zero. As expected the eigenvalue λ decreased to $\lambda = 0.2471$. The decrease in radius was of the order of 3%, hence the effect of convection is relatively important.

Equation (13) was derived based on the fact that the properties of both the liquid and solid phase are constant. In order to evaluate the effect of this assumption, we increased the properties of the liquid phase by 10%. The new root is $\lambda = 0.2545$, which is 0.1% lower from the original root $\lambda = 0.2549$, hence we can claim that our assumption of constant properties has limited impact on the results.

The temperature distribution in the solid powder as a function of the radius r from the cylinder center starting at the liquid-solid interface (i.e., radius of the molten pool of 0.12 mm), as calculated by solving Equation (10), is shown in Figure 5.

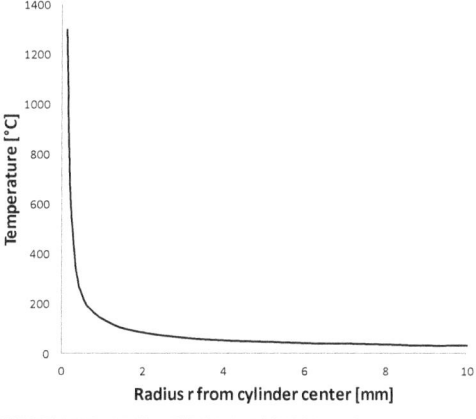

Figure 5. Temperature distribution as a function of radius r from the cylinder center with a melt pool radius of 0.12 mm at time $t = \tau$. The power per unit length is set to $\dot{Q}' = 10000$ W/m.

3.2. Optimum Value of the Line Heat Source

While the heat source is active (for the time period τ), the location of the interface of the melted material is given by $R[t] = 2\lambda\sqrt{\alpha_\ell t}$ for $t \leq \tau$. Of course, melting may continue for a brief period of time after the heat source is deactivated; however, this is beyond the scope of the current work. As expected, the longer the time period τ and the higher the power $\dot{Q}'$, the larger the radius $R[\tau]$. This raises the question on how the power is affecting the radius of melting for a fixed value of energy input per unit length:

$$Q' = \dot{Q}' \cdot \tau \; (\text{J/m}). \qquad (14)$$

Hence, in Equation (13), we replaced $\dot{Q}'$ with Q'/τ, and found the value of the eigenvalue λ as a function of the activation time τ of the heat source. For each value of the eigenvalue we determined the radius of the melted material at time τ (i.e., $R[\tau] = 2\lambda\sqrt{\alpha_\ell \tau}$). As an example, we set the total energy input to $Q' = 150$ (J/m). For the range of τ between 0.0001 and 0.1, we obtained that there was an optimum power per unit length $\dot{Q}'$, such that $R[\tau]$ was maximized, as shown in Figure 6.

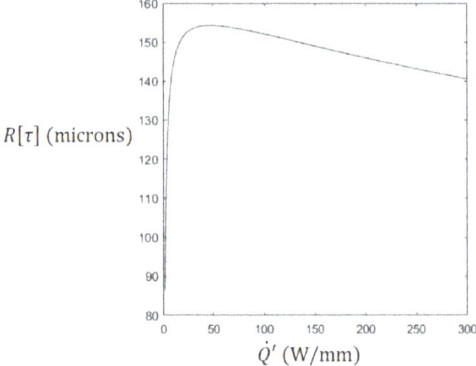

Figure 6. Radius of the melted material at time $t = \tau$, as a function of power per unit length. The energy per unit length is constant $Q' = \dot{Q}' \cdot \tau = 150$ J/m.

From the numerical solution of Equation (13) for a fixed amount of energy per length $Q' = 150$ J/m, we obtained that the optimum power of the line heat source was approximately $\dot{Q}' = 48.4$ kW/m, which would lead to a radius $R[\tau = 0.0031s] = 154$ microns.

However, in real SLM processes such short time periods τ (i.e., velocities in the order of 100 m/s) are not feasible. State-of-the-art processing speeds were found in the literature to be in the order of 0.01 to 2 m/s for high performance metallic alloys with a laser power of 100 to 1000 W.

4. Application of the Model to an SLM Process

Selective laser melting (SLM) is an additive rapid manufacturing technique where a laser is used to fuse metal powder that consists of micro- and nano-particles, into a specified three-dimensional geometry. In this section, we will apply the results of Section 2 (Equation (13)) to experimental data obtained in an SLM (selective laser melting) manufacturing process.

In Tables 1 and 2, we show the results of five experiments for the investigated IN718 alloy, based on the macrographs of a real SLM process, which are shown in Figure 7. The experiments were performed in the context of the MERLIN project on an SLM 280 HL machine (SLM solutions) with varying scan speed and laser powder [13]. Further process parameters were kept constant: laser focus diameter $d_f = 90$ μm (Gauss), hatch distance $\Delta y = 80$ μm, and layer thickness $D = 30$ μm. The weld pool geometry that resulted in each experiment approximates a cylindrical shape with a varying degree of deviation to the cylindrical shape because of surface (radiation, gas convection, Marangoni effect) and depth effects (conduction to solid substrate, gravity etc). Whereas experiment 1, 2, and 4 proved a marginal deviation from the cylindrical shape, a higher deviation was noticed for higher energy per unit length, i.e., experiment 5, due to the higher laser power and lower scan speed. This phenomenon is known as the keyhole effect during which the laser power density is so high that the metallic material reaches temperatures beyond melting, i.e., it vaporizes. The vaporizing metal reaching the gas state expands creating a keyhole or a capillary penetrating from the surface down to weld depth. As the laser beam moves across the surface, the keyhole follows and creates a typically deep and narrow weld. As long as the laser power is great enough and the travel speed is not excessive, this keyhole will remain open. Process parameter combinations which lead to higher energy per unit length values and, thus, higher weld pool penetration, apart from being energy costly and/or slow, are prone to higher sensitivity to porosity [29]. In order to achieve process feasibility in SLM, keyhole effects in melt pools are to be avoided, so that overlap with underlying layers and adjacent scan vectors during processing can be attained without failures. For this reason, a cylindrical approach proves to be consistent for modeling melt pool geometries for SLM applications. As the numerical/analytical results (Equation (13)) were obtained under the assumption of cylindrical symmetry (i.e., the cross-section of the melt pool is a semi-circle), for comparison with the experimental results we introduced the equivalent radius R (Tables 1 and 2) (i.e., the radius of the semicircle with an area equal to the area of experimental melt pool; Figure 7).

Table 1. SLM experiments using the alloy IN718. In the table we show the data of the five experiments, the energy per unit length delivered by the laser, and the upper bound (i.e., the maximum possible radius of molten material).

	Experiment 1	Experiment 2	Experiment 3	Experiment 4	Experiment 5
$\dot{Q}$ (W)	300	360	240	300	300
v (m/s)	1.6	1.6	1.6	1.92	1.28
width (μm)	147	139	110	127	140
depth (μm)	94	83	53	84	139
Q' (J/m)	188	225	150	156	234
Equivalent radius $R = \sqrt{\text{width}/2 \times \text{depth}}$	83	76	54	73	99
Maximum radius (Equation (15))	183	200	164	167	205
Relative error (%)	55	62	67	56	52

Table 2. SLM experiments using the alloy IN718. In the table we show the data of the five experiments, the equivalent heat source, the equivalent radius of the elliptical pool, and the radius calculated using Equation (13).

	Experiment 1	Experiment 2	Experiment 3	Experiment 4	Experiment 5
$\dot{Q}$ (W)	300	360	240	300	300
v (m/s)	1.6	1.6	1.6	1.92	1.28
width (μm)	147	139	110	127	140
depth (μm)	94	83	53	84	139
Q' (J/m)	188	225	150	156	234
$\dot{Q}'_{eq}$ (W/m)	1293	1552	1035	1078	1616
Equivalent radius $R = \sqrt{\text{width}/2 \times \text{depth}}$	83	76	54	73	99
Numerical radius (Equation (13))	70	85	54	52	99
Relative error	15%	12%	<1%	28%	<1%

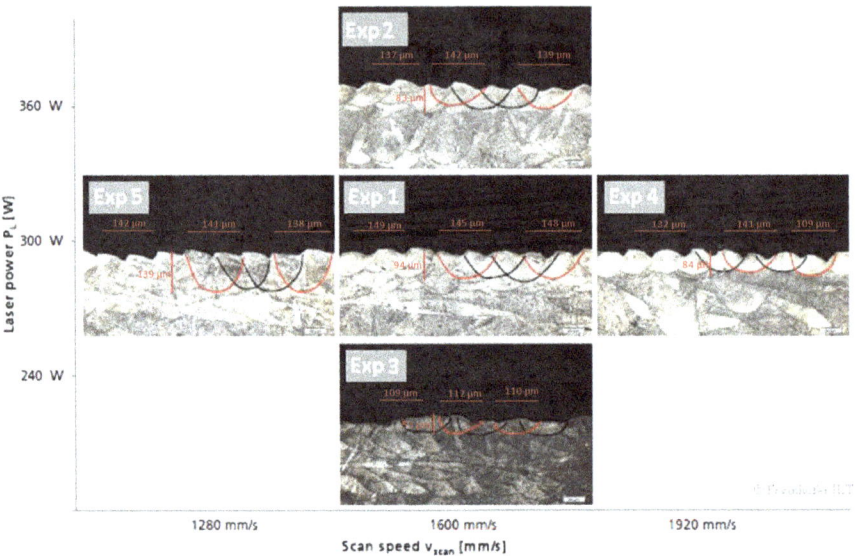

Figure 7. Macrographs of five selective laser melting (SLM) experiments with varying process parameters [13].

An upper bound of the radius of the melt pool was obtained by assuming no conduction, and that all the energy per meter Q' delivered by the laser was used to melt the powder at the melting temperature T_{melt}:

$$Q'(\text{J/m}) = \dot{Q}/v = \rho_s \frac{\pi R^2}{2} c_{ps} \delta T + \rho_s \frac{\pi R^2}{2} L$$
$$\Rightarrow R = \sqrt{\frac{2Q'}{\rho_s \pi (c_{ps} \delta T + L)}} = \sqrt{\frac{2Q'}{3926 \pi (351 \times 1280 + 459360)}} \tag{15}$$

where R represents the equivalent radius, $\dot{Q}$ (W) is the laser power and v(m/s) is the velocity of the laser beam. In Table 1, we show the experimental data associated with the five experiments, the energy per unit meter delivered by the laser Q'(J/m), the equivalent radius, and the maximum possible (upper bound) radius of molten material that can be achieved, as described by Equation (15). It is easily concluded that only a small fraction of the available laser energy is responsible for the melting of the material. Hence, it is expected that a model that includes conduction would provide better results. Such a model is the model developed in Section 2 of the paper.

In order to compare the experimental data (Figure 7) with the results of Section 2, a number of assumptions/parameters were adopted. As already described in Section 2, the metal powder, which rests on a metal substrate, is molten by the scanning laser beam and solidifies into a metal deposit fused to the substrate. The density and the thermal material properties of the molten material during the fusion process are significantly different from the metal powder, as presented in Section 3. Furthermore, we assumed that during the initial stages of melting, the gravity and Marangoni effects can be neglected, hence the process is cylindrically symmetric. Since the conductivity of the metal powder is in the order of 100 times smaller than that of the metal substrate, we assumed that the heat transfer process is controlled by the powder bed, and the heat transfer process is prevalent in the molten pool [1]. Hence, during the initial melting of the powder, we neglected the presence of the metal substrate and assumed a semi-infinite layer of metal powder. The melting is achieved by a rapidly scanning laser beam of power $\dot{Q}$(W) moving with velocity v (m/s) along the span of a powder-bed, and delivering an amount of energy Q'(J/m) = $\dot{Q}/v$ per unit length/span [24]. The energy is conducted through the powder-bed, which was assumed to be a continuum medium. Since the optical or mechanical scanning of the laser beam is much faster than the thermal dynamics in the material, the process can be modeled as a heat source of power $\dot{Q}'(\text{W/m})$ per unit length, distributed simultaneously and evenly along the inner surface $r = R_i[t]$ of the molten material (Figures 1 and 2 [24]), due to irradiation and reflections. As the heat is conducted through, a phase-change takes place. In particular, an interface $r = R[t]$ is developed at temperature T_{melt} that separates the molten material (liquid) from the solid material (powder), where T_{melt} is the melting temperature of the material. Assuming a pure substance, and by ignoring the kinetics of phase change in melting (i.e., the latent heat of fusion L is provided instantaneously), the interface is sharp ($r = R[t]$). Furthermore, in order to relate the experimentally obtained Q' with the $\dot{Q}'$ of the numerical model, we defined the characteristic time t_c and length ℓ_c of the process. The characteristic parameters are process parameters that were obtained experimentally for a set of experiments performed under similar conditions. They can be obtained by fitting the numerical model (Equation (13)) to the available experimental data (Table 2), and can be continuously updated in-process to obtain improved values. In order to find an equivalent power per unit length ($\dot{Q}'_{eq}$) to use in the numerical simulations (Equation (13)) and an activation time (τ), we used the characteristic parameters as follows:

$$\dot{Q}'_{eq}(\text{W/m}) = Q'/t_c = \dot{Q}/(v \times t_c),$$
$$\tau = \ell_c/v,$$

where the energy delivered to the powder bed per unit length is equal to

$$Q'(\text{J/m}) = \dot{Q}/v.$$

Hence, in the characteristic equation for the eigenvalue λ and radius $R[\tau]$ (Equation (13)), we replaced $\dot{Q}'$ with $\dot{Q}'_{eq}$ and τ with ℓ_c/v. Furthermore, the experimentally-determined melt pool is not circular, hence, for comparison with the numerical results, we used the equivalent radius, described earlier.

Using the data of the experiments and Equation (13) we performed a best fit, and estimated a characteristic time $t_c(s) = 0.121$ s and length $\ell_c = 0.099$ m. The numerically calculated radius is shown in the second to last row of Table 2. Out of the five sets of data, we obtained an excellent fit for two sets, a reasonable fit for the other two sets, and a poor fit for one data set among experimental and numerical results as illustrated in Figures 8 and 9. The fact that our model required two fitted parameters, it was expected to fit at least two sets of data well. Although the fitting (Table 2) looks promising, a larger amount of experimental data is necessary in order to justify its applicability to an SLM process. For example, we used a far field temperature of $T_\infty = 20\,^\circ\text{C}$. However, temperature changed during the course of the experiment due to pre-heating, which occurred in the processed layer due the conduction during laser scanning of the adjacent powder (i.e., previously deposited and underlying beads). This was because in actual SLM there is no time for previous deposits and the substrate to cool down to room temperature before the current new bead is processed. Furthermore, the deviation from the circular shape, although it should be avoided for an SLM process, is a dominant effect in the experimental data. Finally, although the power of the laser was known, the total energy delivered to the melt pool was unknown (i.e., the duration of the laser scanning was not reported). Hence, although the results shown in Table 1 prove that the aforementioned model and assumptions could provide reliable results in a very efficient manner in terms of computation time, further controlled experiments are necessary in order to improve, justify, and extend the applicability of the model.

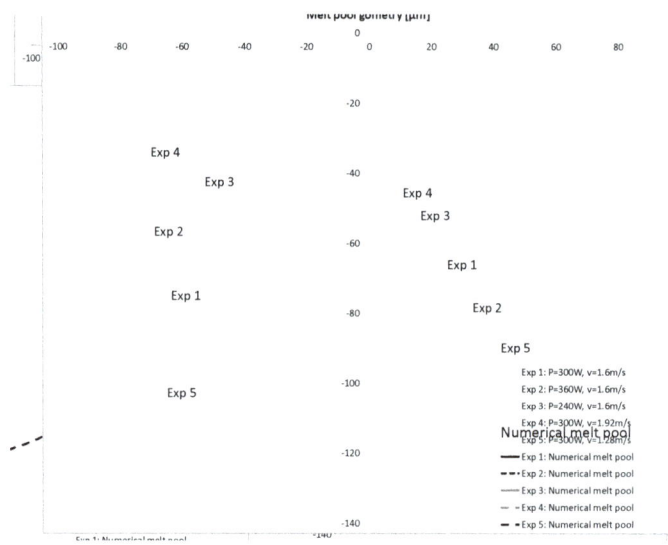

Figure 8. Comparison of the numerical melt pool cylindrical shape (**right**) vs. the measured melt pool dimensions (**left**) for the five sets of experiments.

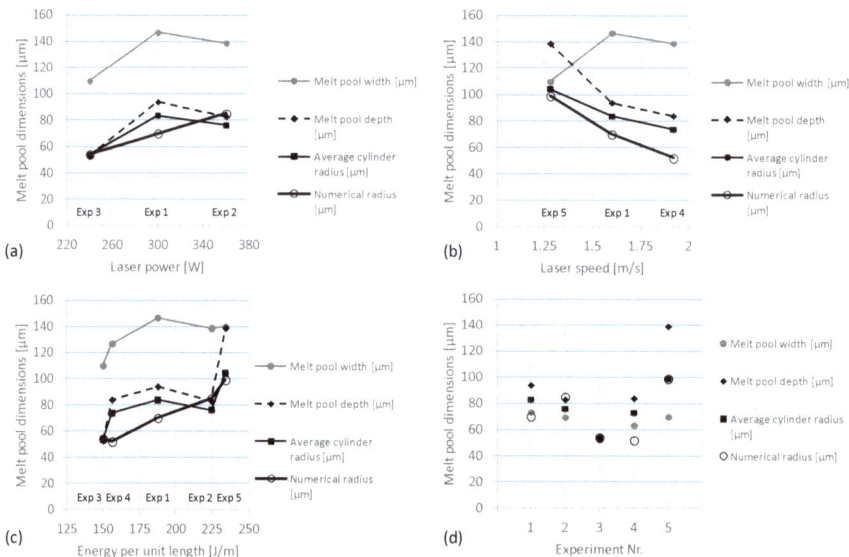

Figure 9. Numerical vs. experimental results of the melt pool dimensions with varying laser power (**a**), laser speed (**b**), and energy per unit length (**c**). The deviation of the proposed analytical model compared to real melt pool dimensions for all five experiments is summarized in (**d**).

A comprehensive illustration of the results in Table 2 is provided in Figure 8. The numerical results of the melt pool cylindrical shape on the right of the figure are compared with the experimental measurements of the melt pool on the left of the figure for all five experiments.

Figure 9 summarizes the numerical and experimental results of the melt pool dimensions in terms of the variation of (a) the laser power, (b) the laser speed, and (c) the energy per unit length. The results for all five experiments (d) demonstrate the deviation of the proposed analytical model compared to real melt pool dimensions.

5. Summary and Conclusions

We address the classical problem of a heat source activated along the span of a semi-infinite material. The material is solid (metal powder) and the melting commences immediately at the location of the line heat source due to the singular nature of the heat input. The temperature profile is cylindrically symmetric and the liquid-solid interface propagates radially outwards. The contribution of the current work is that we have included convection effects due to the density difference between the solid (metal powder) and the liquid (molten powder). Because the liquid has higher density than the powder, an annulus is developed with a radially-symmetric flow field; at small times we neglect gravitational and Marangoni effects. We show analytically that the convection enhances the melting speed and it is relatively important. We also show that if the properties of both the liquid and the solid phases of the material are varied by 10%, that does not affect significantly the results, hence constant properties can be used in numerical simulations. An important result of this work; however, is that, for a fixed amount of input energy per unit length, there exists an optimum combination of power input and time period of activation of the line heat source, which would lead to a maximum radius of the liquid-solid interface (i.e., a maximum amount of material would have melted during the activation time).

Finally, we use the analysis of the heat source per unit length to model an SLM process. We show that the point heat source associated with an SLM process can be represented with an equivalent heat source per unit length using an experimentally-determined characteristic time and length. The benefit of using a heat source model of this kind is twofold. On one hand, a rapid in-process computation of the melt pool can be achieved which can facilitate a real-time process parameter observation and optimization within the feasible process window, and on the other hand it can be applied for process control purposes during the SLM process. Additionally, such a simplified heat source model can be transferred comfortably in a subsequent thermo-mechanical model for computation of the structural shape and residual stresses. A comparison with experimental results has shown that the numerical results could provide a quantitative description of the process. However, further experiments are necessary in order to determine the conditions of applicability of the analytical model.

Author Contributions: Conceptualization, M.M.F., L.P., C.R., A.M. and C.C.D.; Data curation, M.M.F., Y.I. and L.P.; Formal analysis, M.M.F., L.P. and C.C.D.; Funding acquisition, L.P.; Investigation, M.M.F. and L.P.; Methodology, M.M.F., L.P., C.R., A.M. and C.C.D.; Project administration, L.P.; Resources, M.M.F., Y.I. and L.P.; Software, M.M.F.; Supervision, M.M.F., C.R. and C.C.D.; Validation, M.M.F., L.P. and C.C.D.; Visualization, M.M.F., Y.I. and L.P.; Writing—original draft, M.M.F. and L.P.; Writing—review & editing, L.P., C.R. and C.C.D.

Funding: The experimental results presented in this paper were accomplished by Fraunhofer ILT in the framework of the MERLIN Project which has received funding from the European Commission's 7th Framework Programme FP7 2007–2013 under the Grant agreement 266271. Project website: http://www.merlin-project.eu. The author Marios M. Fyrillas was partially supported by CRoNoS—IC408 COST Action.

Conflicts of Interest: The authors declare no conflict of interest.

Nomenclature

c_p	specific heat (J/(kg K))
k	thermal conductivity (W/(m K))
$\dot{Q}$	laser power (W)
$\dot{Q}'$	continuous line heat source (W/m)
Q'	energy per unit length (J/m)
T	temperature (°C)
r	radius of half cylinder (m)
$R_i(t)$	inner radius of melted material where laser heat source is applied (i.e., liquid-air interface) (m)

$R(t)$	outer radius of melted material (i.e., liquid-solid interface)
$\dot{R}(t)$	rate of change of outer radius (i.e., radial velocity) (m/s)
$u_r[r,t]$	radial velocity of the fluid (m/s)
v_ℓ	velocity of the fluid perpendicular to the interface r = R[t] (m/s)
L	latent heat (J/kg)
H	enthalpy (J/kg)
z	normalized radius r/R[t] (-)
c	constant
R	equivalent radius (m)
v	laser scan velocity (m/s)
ℓ_c	characteristic length of scan vector (m)
t_c	characteristic time (s)
d_f	laser focus diameter (m)
D	layer thickness (mm)

Greek symbols

α	thermal diffusivity (m^2/s)
ρ	density (kg/m^3)
τ	time period (s)
ε	ratio $(\rho_\ell - \rho_s)/\rho_\ell$
λ	eigenvalue
Δy	hatch distance (m)

Subscripts

s	solid phase
ℓ	liquid phase

References

1. Carslaw, H.H.; Jaeger, J.C. *Conduction of Heat in Solids*; Oxford University Press: Oxford, UK, 1959; p. 296.
2. Paterson, S. Propagation of a boundary of fusion. *Proc. Glasgow Math. Assoc.* **1952**, *1*, 42–47. [CrossRef]
3. Hu, H.; Argyropoulos, A.S. Mathematical modelling of solidification and melting: A review. *Model. Simul. Mater. Sci. Eng.* **1996**, *4*, 371–396. [CrossRef]
4. Alexiades, V.; Lunardini, V.J.; Solomon, A.D. Mathematical Modeling of Melting and Freezing Processes. *J. Sol. Energy Eng.* **1993**, *115*, 121. [CrossRef]
5. Özişik, M.N. *Heat Conduction*; John Wiley & Sons: Hoboken, NJ, USA, 1980; p. 402.
6. Font, F.; Myers, T.G.; Mitchell, S.L. A mathematical model for nanoparticle melting with density change. *Microfluid. Nanofluidics* **2014**, *18*, 233–243. [CrossRef]
7. Eller, A. Rectified Diffusion during Nonlinear Pulsations of Cavitation Bubbles. *J. Acoust. Soc. Am.* **1965**, *37*, 493–503. [CrossRef]
8. Plesset, M.S.; Zwick, S.A. A Nonsteady Heat Diffusion Problem with Spherical Symmetry. *J. Appl. Phys.* **1952**, *23*, 95. [CrossRef]
9. Brenner, P.B.; Hilgenfeldt, S.; Lohse, D. Single bubble sonoluminescence. *Rev. Mod. Phys.* **2002**, *74*, 425–484. [CrossRef]
10. Chu, S.; Prosperetti, A. History effects on the gas exchange between a bubble and a liquid. *Phys. Rev. Fluids* **2016**, *1*, 1. [CrossRef]
11. Fyrillas, M.M.; Szeri, A.J. Dissolution or growth of soluble spherical oscillating bubbles. *J. Fluid Mech.* **1994**, *277*, 381. [CrossRef]
12. Cheng, B.; Chou, K. Melt Pool Evolution Study in Selective Laser Melting. In *Proceedings of the 26th Annual International Solid Freeform Fabrication Symposium, Austin, TX, USA, 10–12 August 2015*; Bourell, D.L., Ed.; University of Texas: Austin, TX, USA, 2015; pp. 1182–1194.
13. MERLIN Project. *Development of Aero Engine Component Manufacture using Laser Additive Manufacturing, 7th Framework Programme FP7*; Rolls-Royce plc: Derby, UK, 2013.
14. Polivnikova, T. Study and Modelling of the Melt Pool Dynamics during Selective Laser Sintering and Melting. Ph.D. Thesis, Ecole Polytechnique Federale de Lausanne, Lausanne, Switzerland, 2015.

15. Li, Y.; Zhou, K.; Tor, S.B.; Chua, C.K.; Leong, K.F. Heat transfer and phase transition in the selective laser melting process. *Int. J. Heat Mass Transf.* **2017**, *108*, 2408–2416. [CrossRef]
16. Letenneur, M.; Kreitcberg, A.; Brailovski, V. Optimization of Laser Powder Bed Fusion Processing Using a Combination of Melt Pool Modeling and Design of Experiment Approaches: Density Control. *J. Manuf. Mater. Process.* **2019**, *3*, 21. [CrossRef]
17. Li, Y.; Zhou, K.; Tan, P.; Tor, S.B.; Chua, C.K.; Leong, K.F. Modeling temperature and residual stress fields in selective laser melting. *Int. J. Mech. Sci.* **2018**, *136*, 24–35. [CrossRef]
18. Tan, P.; Shen, F.; Li, B.; Zhou, K. A thermo-metallurgical-mechanical model for selective laser melting of Ti6Al4V. *Mater. Des.* **2019**, *168*, 107642. [CrossRef]
19. Doumanidis, C. Simulation for control of sequential and scanned thermal processing. *Int. J. Modeling Simul.* **1997**, *17*, 169–177. [CrossRef]
20. Doumanidis, C.; Fourligkas, N. Temperature Distribution Control in Scanned Thermal Processing of Thin Circular Parts. *IEEE Trans. Control Syst. Technol.* **2001**, *9*, 708–717. [CrossRef]
21. Doumanidis, C.; Hardt, D.E. Simultaneous in-process control of heat-affected zone and cooling rate during arc welding. *Weld. Res. Suppl.* **1990**, *69*, 186–196.
22. Kruth, J.P.; Duflou, J.; Mercelis, P.; Van Vaerenbergh, J.; Craeghs, T.; De Kuester, J. On-line monitoring and process control in selective laser melting and laser cutting. In Proceedings of the 5th Lane Conference, Laser Assisted Net Shape Engineering, Erlangen, Germany, 25–28 September 2007; pp. 23–37.
23. Doumanidis, C.C. Modeling and Control of Timeshared and Scanned Torch Welding. *J. Dyn. Syst. Meas. Control* **1994**, *116*, 387–395. [CrossRef]
24. Papadakis, L.; Loizou, A.; Risse, J.; Bremen, S.; Schrage, J. A Computational Reduction Model for Appraising Structural Effects in Selective Laser Melting Manufacturing: A Methodical Model Reduction Proposed for Time-Efficient Finite Element Analysis of Larger Components in Selective Laser Melting. *J. Virtual Phys. Prototyp.* **2014**, *9*, 17–25. [CrossRef]
25. Doumanidis, C.; Fyrillas, M.; Polychronopoulou, K.; Ioannou, Y. Analytical model for geometrical characteristics control of laser sintered surfaces. *Int. J. Nanomanuf.* **2010**, *6*, 300.
26. Panton, L.P. *Incompressible Flow*; John Wiley & Sons: Hoboken, NJ, USA, 2005.
27. Davis, J.R. *ASM Specialty Handbook: Heat-Resistant Materials*; ASM International: Geauga County, OH, USA, 1997.
28. Hosaeus, H.; Seifter, A.; Kaschnitz, E.; Pottlacher, G. Thermophysical Properties of Solid and liquid Inconel 718. *Alloy. High Temp. High Press.* **2001**, *33*, 405–410. [CrossRef]
29. Shi, X.; Ma, S.; Liu, C.; Chen, C.; Wu, Q.; Chen, X.; Lu, J. Performance of High Layer Thickness in Selective Laser Melting of Ti6Al4V Materials. *Materials* **2016**, *9*, 975. [CrossRef] [PubMed]

© 2019 by the authors. Licensee MDPI, Basel, Switzerland. This article is an open access article distributed under the terms and conditions of the Creative Commons Attribution (CC BY) license (http://creativecommons.org/licenses/by/4.0/).

MDPI
St. Alban-Anlage 66
4052 Basel
Switzerland
Tel. +41 61 683 77 34
Fax +41 61 302 89 18
www.mdpi.com

Journal of Manufacturing and Materials Processing Editorial Office
E-mail: jmmp@mdpi.com
www.mdpi.com/journal/jmmp

www.ingramcontent.com/pod-product-compliance
Lightning Source LLC
LaVergne TN
LVHW072000080526
838202LV00064B/6801